SpringerBriefs in Computer Science

More information about this series at http://www.springer.com/series/10028

Longxiang Gao • Shui Yu • Tom H. Luan
Wanlei Zhou

Delay Tolerant Networks

 Springer

Longxiang Gao
School of Information Technology
Deakin University
Burwood, VIC, Australia

Shui Yu
School of Information Technology
Deakin University
Burwood, VIC, Australia

Tom H. Luan
School of Information Technology
Deakin University
Burwood, VIC, Australia

Wanlei Zhou
School of Information Technology
Deakin University
Burwood, VIC, Australia

ISSN 2191-5768　　　　　　　　ISSN 2191-5776　(electronic)
SpringerBriefs in Computer Science
ISBN 978-3-319-18107-3　　　　ISBN 978-3-319-18108-0　(eBook)
DOI 10.1007/978-3-319-18108-0

Library of Congress Control Number: 2015937970

Springer Cham Heidelberg New York Dordrecht London

Printed on acid-free paper

Springer International Publishing AG Switzerland is part of Springer Science+Business Media (www.springer.com)

Preface

The delay-tolerant networks (DTNs) are emerging research topics that have attracted keen research efforts from both academia and industry. Different from the traditional communication networks, DTNs consider an extreme network condition where a complete end-to-end path between the data source and destination may not exist, and the network is subject to dynamic node connections and unstable topologies. With the above features, DTNs find broad applications in the situations where legacy networks cannot work effectively, such as data communications in rural areas, where stable communications infrastructure is not available or costly, and crucial areas, e.g., disaster rescue and battlefield communications. To summarize, the DTNs, as an important technology complementary to traditional networkings, can be widely applied to national welfare and the people's livelihood.

This monograph reviews the popular and dominant applications and routing protocols in DTNs. Furthermore, new research trends in DTNs are also introduced, such as the exploration of social attributes, time pattern and geographic information for DTN routing, and privacy preserving DTN data transmissions. The monograph is organized as follows. Chapter 1 introduces the characteristics of DTNs and describes the contributions of this monograph. Chapter 2 shows the dominant applications of DTNs. Chapter 3 reviews the state-the-of-art routing protocols in DTNs. Chapter 4 presents two novel DTN routing schemes which explore the multiple dimensional social attributes. Chapter 5 studies the impact of geographic information on the vehicle-based and human-associated delay-tolerant networks, respectively. A real-world dataset is used for the performance evaluation. Chapter 6 studies the privacy preserving in human-associated DTNs, and a secured data forwarding model has been presented. Chapter 7 summarizes this monograph and discusses on the future works.

We would like to thank the editor of this series, Professor Sherman Shen, for his constructive guidance and kind help. Special thanks are also due to the staff at Springer, Melissa Fearon and Jennifer Malat, for their help throughout the publication preparation process.

Burwood, Australia

Longxiang Gao
Shui Yu
Tom H. Luan
Wanlei Zhou

Contents

Chapter 1
Introduction

In this chapter, the backgrounds of Delay Tolerant Networks (DTNs) are introduced first. Hot research questions in DTNs are then formally addressed. Lastly, the organization of this monograph is presented at the end of this chapter.

1.1 Background

1.1.1 Definition and Characteristics of Delay Tolerant Networks

Delay Tolerant Networks (DTNs) are featured with long latency and unstable network topology, where end-to-end delay can be measured in days and routing paths may not exist. These features make traditional routing protocols [1–3] for mobile ad hoc networks unsuitable for DTNs [4, 5]. This is because most routing protocols for ad hoc networks need to set up a complete end-to-end route with at least one available end-to-end path between the source and destination before packet transferring in the traditional mobile ad hoc network. It is typically assumed that in DTNs any link between two nodes is bidirectional and the latency of the network is not a concern. On the other hand, DTNs can be widely applied to challenging networks, such as space communications, mobile sensor networks, and military operations. Characteristics of DTNs are a perfect match in these challenging networks.

Data forwarding in DTNs typically adopts store-and-forward techniques: if there is no connection available at a particular time in these techniques, the source node needs to store and carry the message until the next available node is encountered and these encountered nodes form a candidature set. How to efficiently compare and choose which nodes are the most suitable forwarding targets becomes an important research topic in DTNs, where different forwarding selection algorithms

© The Author(s) 2015
L. Gao et al., *Delay Tolerant Networks*, SpringerBriefs in Computer Science,
DOI 10.1007/978-3-319-18108-0_1

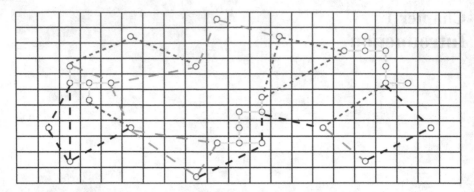

Fig. 1.1 Example of nodes in a geographic topology along with their common social attributes

have different strategies to improve decisions about the routing process [6–8]. Examples of these strategies include how to utilize network resources and balance the congestion level of current network.

On the other hand, when mobile devices in DTNs are associated with human beings, such as a Bluetooth enabled mobile phone carried by a human, they inevitably demonstrate certain social aspects associated with human-to-human communications. More specifically, the human social behaviour determines data communication in human associated DTNs. For example, in Fig. 1.1, the distributed topology of nodes is illustrated by the solid line, where they can physically contact their neighbor nodes. This is similar to traditional routing protocols in MANETs, which mainly focus on analyzing the physical topology movements, such as measuring the distance between two nodes in their geographic dimension. An example of this routing is greedy perimeter stateless routing (GPSR) [9], which has been proven to be an efficient routing scheme when nodes' geographic positions are stable. However, geographic topology in DTNs is unstable, while social characters of mobile nodes are relatively stable for a fixed period of time. Therefore, moving the angle to analyze the social behaviors of each network means the influence of social attributes can be easily detected and this is more important than geographic information in making the route. By using the example in Fig. 1.1, links labelled in dashed lines represent a shared common social attribute, such as the same swimming club membership. This shows that even if there was no physical link, they still can contact each other by their common social habits. A typical example is a Computer Science student would like to communicate with other Computer Science students even though their geographic positions may change from time to time because they may share similar study interests regardless of distance.

Furthermore, these social activities in DTNs are bounded with a time pattern. For example, classmates have a high contact frequency during the daytime, but very low contact frequency during the night. While club waiters may have no contact during the daytime, but a very high contact frequency during the night. This example demonstrates that the time factor may also play an important role in determining the data forwarding performance in DTNs.

1.1.2 Data Forwarding in Delay Tolerant Networks

Traditional data forwarding protocols in MANETs includes two categories: proactive and reactive routings. In proactive routings, forwarding routes are computed automatically and independent of traffic arrivals. Examples of proactive routings are the Destination Sequenced Distance Vector (DSDV) [10] and Optimized Link State Routing (OLSR) [11]. These protocols are capable of computing routes for a connected graph. However, they cannot find a path to link nodes, which are not currently reachable, such as DTNs.

In reactive routings, routes are discovered on demand when traffic needs to be delivered to an unknown destination. Such routing protocols are the Ad-hoc On-demand Distance Vector (AODV) [12] and Dynamic Source Routing (DSR) [13]. In these systems, a route discovery technique is employed to determine a path to the destination. These on-demand routing protocols incur additional delay and they work best when communication patterns are relatively sparse. In delay tolerant networks, as with proactive protocols, these protocols only work for finding routes in a connected subgraph of the overall DTNs routing graph. However, they fail differently to proactive protocols. In particular, they simply fail to return a successful route (from a lack of response), whereas proactive protocols can potentially fail quicker (by determining the requested destination is not presently reachable).

The data forwarding issue is the fundamental problem for delay tolerant networks. To deliver messages in such networks, existing algorithms, such as Epidemic and PRoPHET [14], use flooding or partial flooding with a probability. There is a high probability that any flooding may cause network congestion or high interference, which consumes further resources for processing and switching operations. Context aware routing protocols, such as CAR [15] and HiBOp [16], utilize context information such as history, battery status and changes to the rate of connectivity to compute delivery probabilities. One of the reasons for the above routing mechanisms to use flooding or partial flooding is the lack of routing information. Geographic characters are unreliable and not enough to route messages efficiently due to the nodes' mobility characteristics. However, given the dynamic nature of DTNs, it should take other characteristics into consideration as a viable and beneficial option when we look into ways of increasing efficiency.

1.1.3 Privacy Issue in Delay Tolerant Networks

As the data forwarding issue becomes increasingly important in human involved DTNs, lots of researchers use real social trace files to study the characteristics of mobile nodes. However, these datasets are sensitive, with potential attacks using the provided social attributes to retrieve the subject identification, and therefore the publisher may not want to release these datasets. As a result, how to release high quantity datasets without a privacy issue becomes a critical issue.

Privacy protection is important for researchers. In most cases, researchers use the data anonymous method to pre-process data identifiers and attributes to make sure attackers do not (or take a higher cost to) disclose the dataset with the released data. K-anonymity [17] and Generalizations [18] are typical anonymous methods, which decrease the data utility to protect attribute semantics and therefore prevent the privacy disclosure issue.

Attribute semantics should be reserved in some scenarios, such as data mining or information retrieval. However, it is not necessary in a scenario of DTNs. As the information required for the routing decision making process is the corresponding relationships among attributes, the attributes themselves can be anonymized. If both node's identification and their associated attributes are anonymous, privacy disclosure can be prevented.

1.2   Overview of the Monograph

This section lists and describes the three main topics covered in this monograph: applications based on DTNs techniques, improvement in the performance of data forwarding in DTNs and the use of anonymous data to prevent privacy disclosure in DTNs.

1.2.1   Applications Based on Delay Tolerant Networks

Delay tolerant networks are promising techniques to enable data transmission in challenged scenarios where sophisticated infrastructure is not available and the end-to-end path does not exist at the moment of data transmission. This monograph demonstrates several typical applications based on DTN techniques and studies how to implement them in real-world situations. Applications in this monograph are classified into four categories: digital communication for rural areas, personal/wildlife communications, battlefield communications, disaster rescue and environmental monitoring communications.

1.2.2   Data Forwarding Performance Improvement in Delay Tolerant Networks

How to design an efficient data forwarding model is the main research topic in delay tolerant networks. Two social characteristics based data forwarding models are presented in this monograph. They are the triangle coordinate model and the M-Dimension model. The triangle coordinate model combines the three most useful social attributes into a triangle system, where each angle between two axes represents the social relationship between two social attributes. The smaller an angle is, the closer their social relationship.

The second model, M-Dimension model, provides a general method to combine as many social attributes as possible. These social attributes are stored at a metric and distinguished by measuring the importance of each social attribute. Based on these models, an overall forwarding selection criteria can be established. A multicast based data forwarding algorithm is used to efficiently select inter-node(s) based on this selection criteria to form the concurrent paths from source to destination.

Research in this monograph demonstrates that geographic information plays an important role in both vehicle based and human associated delay tolerant networks. Two geographic delay tolerant network technique based vehicle data dissemination applications are presented. Furthermore, a more comprehensive model in human associated delay tolerant networks is introduced, which takes social attributes, time patterns and geographic information into consideration. With this model, the similarity of each social attribute with time indexed is calculated and the trend for the geographic movements of nodes is examined with a time pattern. An overall social distance from the source to the destination is generated and multicast routing is performed.

1.2.3 Privacy Protected Data Forwarding in Delay Tolerant Networks

As more and more researchers are using real social trace files to study the data forwarding issue in DTNs, the data privacy becomes a serious concern for the data publisher. An entropy based anonymous data detection and forwarding model is presented in this monograph, which is used to automatically detect the certainty of each attribute. For example, if the entropy value of an attribute is larger, it indicates that this attribute has a higher uncertainty and is therefore more important. This entropy based algorithm does not require to disclose the meanings of attributes, and therefore all attributes can be anonymous. Consequently, track provider can safely publish their dataset without violating privacy. Assume there is a small dataset has 10 subjects with equally distribution between 8 different universities and they are either lecturers or professors, where the certainty of the university location attribute is 0.125 and the certainty of the affiliation attribute is 0.5. Therefore, the university location is more important than the affliction as it has a higher chance of two people belonging to the same affiliation category, and they may have lower chance of knowing each other. However, if two people belong to the same university, they have a higher chance of knowing each other. In this instance, a higher weight factor is assigned to the university location attribute.

A socially aware K-core algorithm (S-kcore) is introduced to find the key node based on social information, which is a modification of the K-core algorithm.

By using S-kcore with combined weighting factors, key nodes in the networks can be selected based on their social relationship instead of geographic information, as the location information is restricted because of the privacy issue.

1.3   Organization of the Monograph

The rest of this monograph is organized as follows.

Chapter 2 describes the applications of delay tolerant networks and their advantages. Applications of DTNs are combined into four categories: digital communications for rural areas, personal/wildlife communications, battlefield communications and disaster rescue and environment monitoring communications.

Chapter 3 reviews the existing routing protocols for delay tolerant networks, including epidemic based routing protocols, probability based routing protocols, geographic based routing protocols, social concept based routing protocols and time related routing protocols.

Chapter 4 introduces two data forwarding models in delay tolerant networks, and both of them use multiple social attributes to improve their routing performance. The Triangle Coordinate model uses the three most popular social attributes to form a triangle system. Angles among each of the social attributes represent their relationship and social importance. The M-Dimension model combines multiple-dimensional social attributes, and is a more general model that includes all social attributes. These social attributes are unified and weighted, which can be used to conduct an overall social distance between nodes. Furthermore, the Infocom 06 dataset is introduced and analysis of its social behaviors are presented.

Chapter 5 presents how geographic information can improve the data dissemination performance in both vehicle based and human associated delay tolerant networks. Two geographic based applications are illustrated to show the characteristics of vehicle based delay tolerant networks, where packets are always forwarded to the node which has a shorter geographic distance to the destination compared with itself. Furthermore, a time-varying social contact topology model is presented to study the routing problem in a more challenging human associated delay tolerant network environment. This model considers and includes the time pattern, social attributes and geographic information. Another benchmark dataset in delay tolerant networks, MIT Reality dataset, is also introduced in this chapter.

Chapter 6 studies the privacy protection issue in DTNs. An Entropy based anonymous attributes detection method and an improved social-aware K-core algorithm are used to obtain the most popular node without disclosing the node's identifier and attribute semantics. An unicast forwarding algorithm is deployed to test the routing performance in a single copy forwarding environment.

Chapter 7 concludes the whole monograph, summarizes the main contributions of this monograph and presents future works.

References

1. Sarkar, N.I., Lol, W.G.: A study of MANET routing protocols: joint node density, packet length and mobility. In: IEEE Symposium on Computers and Communications, Riccione, pp. 515–520 (2010)
2. Kum, D.-W., Park, J.-S., Cho, Y.-Z., Cheon, B.-Y.: Performance evaluation of AODV and DYMO routing protocols in manet. In: Proceedings of IEEE CCNC, Las Vegas, pp. 1046–1047 (2010)
3. Mittal, S., Kaur, P.: Performance comparison of AODV, DSR and ZRP routing protocols in MANET'S. In: International Conference on Advances in Computing, Control, and Telecommunication Technologies, Trivandrum, pp. 165–168 (2009)
4. Samuel, H., Zhuang, W., Preiss, B.: DTN based dominating set routing technique for Mobile Ad Hoc Networks. In: Proceedings of ICST Conference on Heterogeneous Networking for Quality, Reliability, Security and Robustness, Hong Kong, pp. 9:1–9:7 (2008)
5. Samuel, H., Zhuang, W., Preiss, B.R.: Improving the dominating-set routing over delay-tolerant mobile ad-hoc networks via estimating node intermeeting times. EURASIP J. Wireless Commun. Netw. pp. 7:1–7:12 (2011)
6. Zhu, H., Lin, X., Lu, R., Fan, Y., Shen, X.: SMART: A secure multilayer credit-based incentive scheme for delay-tolerant networks. IEEE Trans. Veh. Technol. **58**, 4628–4639 (2009)
7. Samuel, H., Zhuang, W.: Preventing unauthorized messages and achieving end-to-end security in delay tolerant heterogeneous wireless networks. JCM **5**(2), 152–163 (2010)
8. Zhu, H., Du, S., Gao, Z., Dong, M., Cao, Z.: A probabilistic misbehavior detection scheme toward efficient trust establishment in delay-tolerant networks. IEEE Trans. Parallel Distrib. Syst. **25**(1), 22–32 (2014)
9. Karp, B., Kung, H.T.: GPSR: Greedy perimeter stateless routing for wireless networks. In: MobiCom '00: Proceedings of the 6th Annual International Conference on Mobile Computing and Networking, Boston, pp. 243–254. ACM, New York (2000)
10. Perkins, C.E., Bhagwat, P.: Highly dynamic destination-sequenced distance-vector routing (DSDV) for mobile computers. In: Proceedings of the Conference on Communications Architectures, Protocols and Applications, London, pp. 234–244 (1994)
11. Dhurandher, S., Obaidat, M., Gupta, M.: A reactive optimized link state routing protocol for mobile ad hoc networks. In: Proceedings of IEEE ICECS, Athens (2010)
12. Perkins, C., Royer, E.: Ad-hoc on-demand distance vector routing. In: Proceedings of IEEE WMCSA, New Orleans, pp. 90–100 (1999)
13. Wu, J.: An extended dynamic source routing scheme in ad hoc wireless networks. In: Proceedings of Annual Hawaii International Conference on System Sciences, Big Island, pp. 3832–3838 (2002)
14. Lindgren, A., Doria, A., Schelén, O.: Probabilistic routing in intermittently connected networks. SIGMOBILE Mobile Comput. Commun. Rev. **7**, 19–20 (2003)
15. Musolesi, M., Mascolo, C.: CAR: Context-aware adaptive routing for delay-tolerant mobile networks. IEEE Trans. Mobile Comput. **8**, 246–260 (2009)
16. Boldrini, C., Conti, M., Jacopini, J., Passarella, A.: HiBOp: a history based routing protocol for opportunistic networks. In: IEEE International Symposium on a World of Wireless, Mobile and Multimedia Networks, 2007, Helsinki (WoWMoM 2007), pp. 1, 12, 18–21, June 2007
17. Sweney, L.: Achieving k-anonymity privacy protection using generalization and suppression. Int. J. Uncertain. Fuzziness Knowl. Based Syst. **10**(5), 571–588 (2002)
18. Iyengar, V.S.: Transforming data to satisfy privacy constraints. In: Proceedings of ACM SIGKDD, Edmonton, pp. 279–288 (2002)

Chapter 2
Delay Tolerant Networks Based Applications

This chapter surveys the typical applications of delay tolerant network techniques from four aspects: digital communication for rural areas, personal/wildlife communications, battlefield communications, disaster rescues and environmental monitoring communications.

2.1 Digital Communication for Rural Areas

2.1.1 DakNet

DakNet [1] was developed by researchers from the MIT Media Lab. It provides an asynchronous digital connection on the infrastructure of ad hoc networks, as shown in Fig. 2.1. It has been deployed in remote parts of Cambodia and India at a cost two orders of magnitude less compared to traditional landlines networks.

DakNet uses existing communications and transport infrastructure to form a delay tolerant network, where it combines physical transportation with a wireless data transferral function to extend Internet connectivity to a central hub, such as cyberspace or a post office. DakNet does not need to relay data over a long distance, which is an advantage because it would reduce costs and save a lot of power. Instead, DakNet transmits data over short point-to-point links between kiosks and portable storage devices called mobile access points (MAPs). A MAP can exchange data among public kiosks, Internet enabled devices and non-Internet accessed hubs by using mobile generators mounted on and powered by a bus or motorcycle, as shown in Fig. 2.1.

The point-to-point data transfers use high bandwidth with low-cost WiFi radio transceivers. When a MAP-equipped vehicle comes within range of a

© The Author(s) 2015

L. Gao et al., *Delay Tolerant Networks*, SpringerBriefs in Computer Science,
DOI 10.1007/978-3-319-18108-0_2

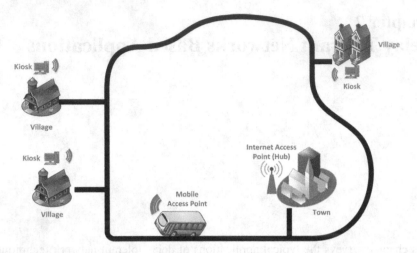

Fig. 2.1 DakNet topology

WiFi-equipped kiosk, it automatically detects the wireless connection and uploads/downloads tens of megabytes of data. On the other hand, when the MAP-equipped vehicle comes within range of an Internet access point, it automatically synchronizes the data generated from rural kiosks. This application creates a low-cost delay tolerant network and asynchronous data exchange infrastructure. This process would be sufficient to provide the daily information services for a small village with only one small vehicle passing everyday.

2.1.2   TrainNet

Zarafshan-Araki and Chin [2] proposed TrainNet, a mechanical backhaul based train to deliver non real-time data using the DTN technique, which provides a low cost and high bandwidth link. The reason to use the train instead of other transportation methods is its fixed timetable where movements are deterministic and cover long distance. It can also carry more storage devices to store more data. The potential application of TrainNet intends to deliver a large non real-time video among any city as long as it has a train system.

Trains and stations in TrainNet need to be equipped with a rack of storages, such as a hard disk, to store, carry and forward data. Each station has a point of presence (POP) to transfer their non real-time data from the provider to the train. As illustrated in Fig. 2.2, large non real-time videos are copped into hard disks by using POP from Rail Station A and put into a train towards to Rail Station B. Once this train arrives, these hard disks are unloaded to the local video service provider through POP. By using this technique, it can save lots of time compared with transferring data through broadband backbone directly.

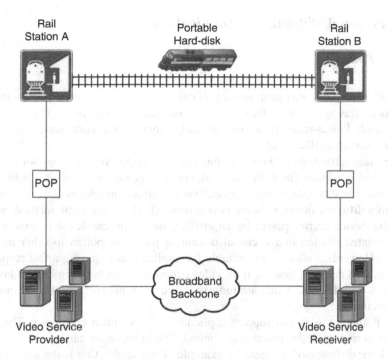

Fig. 2.2 An application example based on TrainNet

2.1.3 KioskNet

Kroeker et al. [3] further explored the importance of establishing a delay tolerant network based on rural kiosks. First of all, rural kiosks play an important role in developing countries. Instead of a pure news agency, rural kiosks have a functional role providing land records, and birth certificates etc. However, without a reliable Internet connection, kiosks are unproductive. Traditional ways of connecting kiosks primarily rely on dial up lines, the satellite terminals or long range WiFi, where they all suffer some bottleneck. For example the dial up line is slow and unreliable, satellite terminal is too expensive and the long range WiFi needs large scale deployment.

KioskNet extends the idea from DakNet [1] to have a complete system with naming, addressing, forwarding and routing. In this system, a kiosk controller is used to provide network boot, storage and user management functions. For those public users, they use the kiosk controller to access the executed applications. For organization users, such as government officials and NGO partners, they treat the kiosk controller as a wireless hotspot to provide a DTN function (store-and-forward) to the Internet. The main advantage of this work is the use of mechanical backhaul as ferries to carry data to the Internet gateway as they pass a kiosk. The mechanical backhaul can be cars, buses, motorcycles or trains. During a one way communication period (20s to 5min), a 100 KB–50 MB data can be transferred.

2.2   Personal/Wildlife Communications

2.2.1   Pollen

The Pollen network was proposed by Natalie et al. in [4]. Every person in the network carries an electronic form of pollen, such as a mobile phone. These devices can be made Pollen-ready by affixing suitably programmed autonomous miniature computers, such as iButtons.[1]

The main advantage of Pollen is the data exchange process does not rely on network infrastructure.The Pollen network borrows the idea that a plant is pollinated by insects. For example, when an insect visits a flower to take nectar, it needs to move to a different flower as some pollen rubs off. Using the same method, when a mobile device carrier passes by an artifact, he or she can leave comments on the associated iButton and it can also transfer pieces of pollen to other mobile devices. However, Pollen is not suitable for applications requiring a fast response or guaranteed delivery. Instead, it provides communication between large numbers of objects and devices, which are too expensive to be networked with the current infrastructure.

The Pollen network can support applications in two main categories. The first category is to provide information in context. The information can benefit from the network for diffusion of updates. An example of this application is the item review system, where reviews can be directly attached to items, such as books and CDs. People can add a review and read other people's reviews. The second category is the aggregate network interactions that can achieve global objectives, such as tracking the location of objects. The Pollen network allows these kinds of applications to become better embedded in a user friendly environments.

2.2.2   Body Area Networks

Quwaider and Biswas [5] introduced the delay tolerant network technique to the Wireless Body Area Network (WBAN) [6, 7] to solve on body packet routing issues in the presence of topological partitioning. Sensor nodes in WBAN are based on low-power supplied RF transceivers [8, 9] where the signal transmission can be affected by postural body movements and clothing.

They developed an on body store and flood routing algorithm to deliver the lower-bound delay in the absence of any delays caused due to packet congestions. This routing algorithm is flooding based, where packets are sent from the source to the

---

[1]iButtons, manufactured by Dallas Semiconductor (now a subsidiary of Maxim Integrated Products), are tiny computers encased in rugged steel containing approximately the size of a large lithium battery.

destination through multiple routes and the first copy received by the destination indicates the minimum possible end-to-end storage delay. The target of this routing algorithm is to minimize the end-to-end delay while ensuring low storage delay.

2.2.3 ZebraNet

Juang et al. [10] designed ZebraNet to track wildlife in Kenya. Costumed tracking collars carried by animals are used to collect a zebra's mobility pattern and send the pattern back to a mobile base station.

As the network between zebras and base stations is opportunistic, the DTN technique is used for this system. By using the DTN technique, ZebraNet stores and forwards collected mobility patterns and receives updated information from the mobile base station, which may not be available at a particular time. In addition, when base stations receive data, there may be no Internet connection, and the data has to remain at the base station. Traditionally, data is collected manually only when researchers visit the field and upload the data to the Internet. This is not efficient as the goal of ZebraNet is to send collected data from each collar back to the base station because the collar may not always be within range of the base station, and therefore, data cannot be transferred directly. Instead, a history-based routing protocol is used to forward data to other collars as intermediate storage. In this routing protocol, when scan for neighbors, if this node is within the range of the base station, this node sends data to base station and increase hierarchy level. Otherwise, it checks hierarchy levels of neighbors and sends data to the neighbor with the highest level and break ties randomly. Eventually the data reaches the base station, where the higher the hierarchy level is, the higher the probability the node is within range of a base station.

2.3 Battlefield Communications

2.3.1 DTN for Military Missions

Free space optical communications (FSOC) [11] are used in battlefield communication to support military missions and provide high capacity bandwidth. However, there are many drawbacks associated with FSOC, as highlighted by Nichols et al. in [12]. Furthermore, fog, dust and atmospheric turbulence affect the performance of RF and any optical transmission. Therefore, a local area communication system is highly demanding and delay tolerant network techniques can fulfil this requirement. Nichols et al. [13] developed an DTN based algorithm to overcome these drawbacks, where they use hop-by-hop decisions based on the temporal frame construct. It helps to move information closer to its destination on each of transactions by considering the topology control requirement from FSOC.

2.3.2 DTN for Airborne Networks

An Airborne Network (AN) is defined as "an infrastructure that provides communication transport services through at least one node that is on a platform capable
of flight" [14]. This includes bombers, helicopters, flight jets and unmanned aerial
vehicles (UAVs). An AN becomes more and more important for the Department of
Defense (DoD) because it provides services for voice commands, mission updates,
tactical video, weather information and so on. Jonson et al. [15] proposed two AN
applications based on the DTN technique.

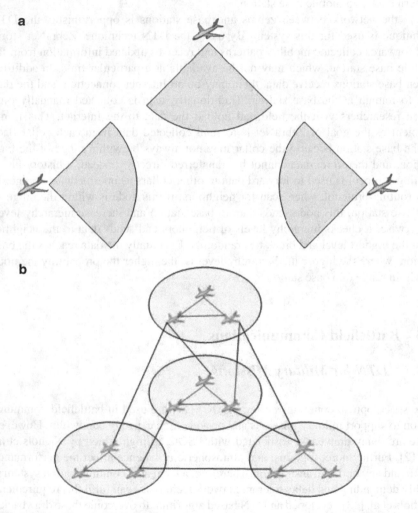

Fig. 2.3 DTN application for airborne network. (**a**) Traditional analogue radio broadcast voice
communication model. (**b**) DTN for airborne networks

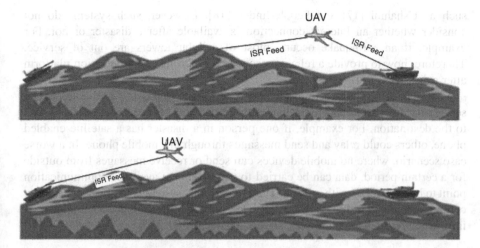

Fig. 2.4 Tactical relay between brigades

The first application is illustrated in Fig. 2.3. Figure 2.3(a) describes an original analogue radio broadcasts voice communications to other planes, where planes should be within a certain radio range with each other. Figure 2.3(b) illustrates a plane acting as a gateway to store and forward data and voice to other planes based on the DTN technique by using low cost digital communication link with short transmission range, such as Tactical Digital Information Links (TADILs).

In the second application, a low orbiting airborne platform, such as a unmanned aerial vehicle (UAV), is used as a media to link two brigades together when they are far away and have been commanded to converge on the same target. As shown in Fig. 2.4, two brigades are unable to set up a communication link to exchange data and video for the same target. Once a UAV is called to provide air support, it collects the data and mission from the first brigade, then flies to the second brigade to update its information and collects the new data and mission.

2.4   Disaster Rescue and Environment Monitoring Communications

2.4.1   DTN for Disaster Response Communications

When a disaster occurs, people inside the affected area need to report the situational awareness (SA),[2] not only for themselves, but also for the emergency rescue [16]. Some online crisis report and management systems have already been developed,

---

[2]This includes both personal information and disaster situations.

such as Ushahidi [17] and People-finder [18]. However, such systems do not consider whether an Internet connection is available after a disaster or not. For example, if an earthquake occurs, most of cellular towers are out of service. Therefore, how to provide a reliable rescue and reporting communication platform after a disaster is a critical issue.

Fall et al. [19] proposed to use the DTN technique to provide a reliable backup service for Internet connectivity that may be the only choice of transferring data to the destination. For example, if one person in a disaster has a satellite enabled phone, others could relay and send messages through this mobile phone. In a worse case scenario, where no mobile devices can send or receive messages from outside for a certain period, data can be carried towards the next available communication point to be exchanged with outside. The target of this system is to provide a reliable communication platform between mobile devices when public communication infrastructure, such as cell towers, are not available.

2.4.2   Sensor Network with Delay Tolerance

SeNDT (Sensor Network with Delay Tolerance), was developed at Trinity College Dublin [20] and is a robust sensor platform that can be deployed for long periods in harsh environments. Two main applications were developed based on this platform: water quality monitoring and environmental noise monitoring.

The water quality monitoring sensor network was used to monitor the water quality of Irish lakes. Traditional water quality monitoring methods take periodic samples at specific locations and analyse them at laboratories. However, this method is not quick enough to efficiently detect phosphates and nitrates on time. After the SeNDT was deployed, water quality samples were collected once or twice per hour compared with once every few weeks with the traditional manual sample collection method.

Another application based on the SeNDT was the noise level monitoring system for Dublin City Council. The purpose was to generate an environmental noise map and combine with population distribution information to determine how many people were affected by the certain level of noise. These sensors were deployed at two urban sites selected by Dublin City Council and main motorways around Ireland selected by the National Roads Authority. The DTN technique was used during the noise data collection process and the collected noise data demonstrated this system worked well.

References

1. Pentland, A., Fletcher, R., Hasson, A.: DakNet: rethinking connectivity in developing nations. Computer **37**(1), 78–83 (2004)

2. Zarafshan-Araki, M., Chin, K.-W.: TrainNet: a transport system for delivering non real-time data. Comput. Commun. **33**, 1850–1863 (2010)
3. Seth, A., Kroeker, D., Zaharia, M., Guo, S., Keshav, S.: Low-cost communication for rural internet kiosks using mechanical backhaul. In: Proceedings of the 12th Annual International Conference on Mobile Computing and Networking, Los Angeles (MobiCom '06), pp. 334–345. ACM, New York (2006)
4. Glance, N., Snowdon, D., Meunier, J.-L.: Pollen: using people as a communication medium. Comput. Netw. **35**, 429–442 (2001)
5. Quwaider, M., Biswas, S.: DTN routing in body sensor networks with dynamic postural partitioning. Ad Hoc Netw. **8**, 824–841 (2010)
6. Qiao, J., Cai, L.X., Shen, X., Mark, J.W.: Enabling multi-hop concurrent transmissions in 60 ghz wireless personal area networks. IEEE Trans. Wireless Commun. **10**(11), 3824–3833 (2011)
7. Jovanov, E., Milenkovic, A., Otto, C., de Groen, P.C.: A wireless body area network of intelligent motion sensors for computer assisted physical rehabilitation. J. Neuroeng. Rehabil. **2**, 6+ (2005)
8. Qiao, J., Shen, X.S., Mark, J.W., Shen, Q., He, Y., Lei, L.: Enabling device-to-device communications in millimeter-wave 5g cellular networks. IEEE Commun. Mag. **53**(1), 209–215 (2015)
9. Cheng, M., Ye, Q., Cai, L.: Rate-adaptive concurrent transmission scheduling schemes for wpans with directional antennas. IEEE Trans. Veh. Technol. (2014)
10. Juang, P., Oki, H., Wang, Y., Martonosi, M., Peh, L.S., Rubenstein, D.: Energy-efficient computing for wildlife tracking: design tradeoffs and early experiences with zebranet. SIGPLAN Not. **37**, 96–107 (2002)
11. Manzur, T.: Free space optical communications (FSO). In: Avionics, Fiber-Optics and Photonics Technology Conference, 2007 IEEE, Victoria, pp. 21–21 (2007)
12. Nichols, R., Hammons, A.: Performance of dtn-based free-space optical networks with mobility. In: Proceedings of IEEE MILCOM, Orlando, pp. 1–6 (2007)
13. Nichols, R., Hammons, A., Tebben, D., Dwivedi, A.: Delay tolerant networking for free-space optical communication systems. In: Sarnoff Symposium, 2007 IEEE, Princeton, pp. 1–5, (2007)
14. Chruscicki, M.C., Hall, F.: Airborne networking component architecture and simulation environment (AN-CASE). In: Proceedings of IEEE MILCOM, Orlando, pp. 1–7 (2007)
15. Jonson, T., Pezeshki, J., Chao, V., Smith, K., Fazio, J.: Application of delay tolerant networking (DTN) in airborne networks. In: Proceedings of IEEE MILCOM, Orlando, pp. 1–7 (2008)
16. Palen, L., Hiltz, S.R., Liu, S.B., Online forums supporting grassroots participation in emergency preparedness and response. Commun. ACM **50**, 54–58 (2007)
17. Liu, S., Palen, L., Sutton, J., Hughes, A., Vieweg, S.: In Search of the Bigger Picture: The Emergent Role of On-Line Photo-Sharing in Times of Disaster. In: Proceedings of the Information Systems for Crisis Response and Management Conference (ISCRAM 2008), Washington, DC (2008)
18. Hughes, A.L., Palen, L.: Twitter Adoption and Use in Mass Convergence and Emergency Events. In: ISCRAM Conference, Gothenburg (2009)
19. Fall, K., Iannaccone, G., Kannan, J., Silveira, F., Taft, N.: A disruption-tolerant architecture for secure and efficient disaster response communications. In: Proceedings of ISCRAM, Seattle, 2010.
20. McDonald, P., Geraghty, D., Humphreys, I., Farrell, S., Cahill, V.: Sensor network with delay tolerance (sendt). In: Proceedings of IEEE ICCCN, Honolulu, pp. 1333–1338 (2007)

Chapter 3
Routing Protocols in Delay Tolerant Networks

In this chapter, typical routing protocols in delay tolerant networks are presented. These routing protocols are categorized into epidemic based routing protocols, probability based routing protocols, geographic based routing protocols, social concept based routing protocols and time related routing protocols.

3.1 Epidemic Based Routing Protocols

3.1.1 Pure Epidemic

Vahdat and Becker proposed Epidemic Routing [5] in delay tolerant networks, which relies on the theory of epidemic algorithms [6]. This routing protocol uses pair-wise information of messages between nodes to deliver messages to their destination. By using this method, messages are quickly distributed through connected portions of the network. Through this transitive transmission of data, messages have a high probability reaching their destination. To have a better understanding of this protocol, Figs. 3.1 and 3.2 are used to demonstrate how epidemic routing works. The scenario at t_1 shows the source (node S) wants to send a message to the destination (node D), but node S only has direct contact with its neighbour, node C_1, and no connection with the destination and node C_2. Node C_1 store and carry the message from node S to wait a next available node. At a later stage, t_2, node C_1 moves to the area of node C_2 and forwards the message to node C_2. Eventually, node D receives this message through node C_2.

In the epidemic routing protocol, if a path to the destination is not available, messages are buffed at the hosts. A summary vector is used for each of the nodes to collect their own information and exchange it with other nodes when they make contact. After this exchange, each node detects whether other nodes have messages they did not receive before. If this situation occurs, the node updates its summary

© The Author(s) 2015
L. Gao et al., *Delay Tolerant Networks*, SpringerBriefs in Computer Science,
DOI 10.1007/978-3-319-18108-0_3

vector to include new messages. Therefore, if the buffer space is big enough, messages can spread like an epidemic to infect other nodes. However, a global ID is required to assign to each of messages. This is because it needs to determine whether a certain message has been stored or not.

3.1.2   Epidemic with TTL and TTS

Pure Epidemic routing [14] has a high delivery success ratio, however, it does create a significant traffic load and consumes many network resources. Harras et al. [7] used time metrics, such as Time to Live (TTL) and Time to Send (TTS), to limit this flooding process. TTL was used to determine how long the message could be alive in the network before being discarded. TTS is used to determine the waiting period before a message is sent. For example, if a node's disconnection time is larger than the TTS, the message would be directly dropped when received.

Furthermore, Harras et al. [7] introduced a Passive Cure concept. The idea behind this concept is that once a forwarding node or destination node receives a message, it generates a Passive Cure to inform nodes in the network to prevent future retransmissions.

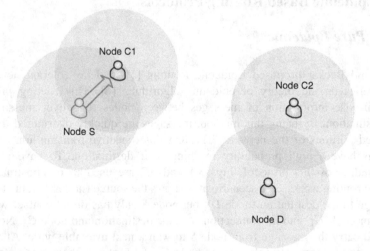

Fig. 3.1 Epidemic routing scenario at t_1

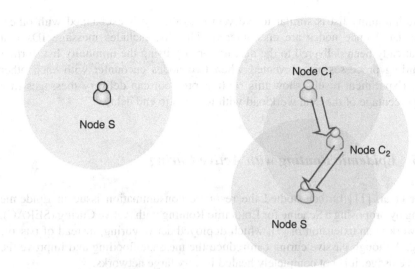

Fig. 3.2 Epidemic routing scenario at t_2

3.1.3 Epidemic with DOA and DLE

Traditional Epidemic routings [8, 14] did not consider the memory problem, where it treats resources unbounded. However, mobile nodes may have limited resource, such as a small buffer size. Therefore, the assumption of an infinite buffer size is not suitable for a general network model. Davis et al. [9] extended Epidemic routing by taking the buffer size into consideration.

Their work adopted two drop strategies: Drop Oldest (DOA) and Drop Least Encountered (DLE). DOA is a strategy to drop the longest packet in the network because if a packet has the longest time in the network, it has most likely been delivered. The second strategy, DLE, is based on the estimated delivery ability. DLE is a history based analysis method to predict the future encountered trend. Experiment results show the DLE is better than DOA.

3.1.4 Epidemic with Immunity

Some popular inter-nodes in DTN are selected to store and forward multiple copies based on Epidemic routing to achieve a higher delivery success ratio, even though its end-to-end delay is high. Mundur et al. [10] introduced the immunity concept, where an immunity-list is exchanged whenever nodes are encountered, to prevent those duplicated store and forward processes. By using this technique, performances of storage utilization are improved and delay is decreased as well.

The immunity list is similar to the vector as the list is exchanged with others and updated once nodes are encountered. This list includes message IDs that have already been delivered to the destination. By using the immunity list, further forwarding processes are prevented when two nodes encounter with each other again. Experiment results show this routing protocol can delivery messages in a high percentage of the total workload with low end to end delay.

3.1.5 Epidemic Routing with Active Curing

Tower et al. [11] further studied the resource consummation issue in Epidemic routing by proposing a Scheme for Epidemic Routing with Active Curing (SERAC). This work is an extension of [7], which deployed active curing, instead of passive curing. Although passive curing can reduce the message flooding and improve the resource usage, it is not completely healed in very large networks.

The main contribution of the work is to make cures (ACK messages) propagate in a more aggressive way. This is much faster than the original infection process with limited resource consumption. Low-power radio based applications can be conducted using this DTN routing protocol.

3.1.6 Spray and Wait

Spyropoulos et al. [12] proposed the Spray and Wait routing protocol. Spray and Wait takes advantage of Epidemic routing [5] with its faster transmission and higher delivery success ratio, which also deployed a direct unicast forwarding process. Spray and Wait has two steps: the Spray step and the Wait step. In the Spray step, a multi-cast process is performed to send multiple copies from a source to the inter-nodes. In the Wait step, inter-nodes, whom receive copies during the Spray Step, use the uni-cast to directly transfer a single copy to its destination. Actually, Spray and Wait routing is a trade off between multi-cast and uni-cast forwarding.

According to other work did by Spyropoulos et al. [13], the upper bound of the expected delay using Spray and Wait routing (ED_{sw}) with M nodes, and L messages in transmission range K is

$$ED_{sw} \leqslant (H_{M-1} - H_{M-L})ED_{dt} + \frac{M-L}{M-1}\frac{ED_{dt}}{L},$$

where H_n is the nth Harmonic Number [14] and ED_{dt} is the expected delay of direct transmission

$$ED_{dt} = 0.5N(0.34logN) - \frac{2^{K+1} - K - 2}{2^K - 1}.$$

Therefore to get the minimum number of messages L_{min} to be transmitted, it needs to make $ED_{sw} = \alpha ED_{opt}$, where $\alpha > 1$ and ED_{opt} is the optimized expected delay, which follows

$$ED_{opt} = \frac{H_{M-1}}{M-1} ED_{dt}.$$

Furthermore, Spyropoulos et al. introduced a binary Spray and Wait method. In the Spray step, every node only forwards $\lfloor n/2 \rfloor$ messages, where n is the total number of message copies and $n > 1$, to its encountered nodes and keeps the rest of the copies. When only one copy is left, this node uses the Wait Step to perform direct transmission.

3.2 Probability Based Routing Protocols

3.2.1 Prioritized Epidemic Routing Protocol

Ramanathan [15] introduced PRioritized EPidemic (PRER) routing to choose the most drapable nodes to reduce the resource usage without affecting the delivery success ratio. Its selection criteria has four parameters: the current cost to destination, the current cost from source, the packet expiry time and the packet generation time.

Instead of keeping the storage and bandwidth utilization at a low level, PREP utilizes these resources by dropping only when necessary. PREP has two modulars: independent components and priority scheme. The independent component estimates the routing cost from a given node to the destination based on the overall topology view. The priority scheme includes the deleting and transmitting process. The deleting process is used whenever the buffer usage exceeds the pre-defined threshold and the transmission process is used whenever there is no transmission appearing in the last T seconds.

3.2.2 Meeting-Visit Routing Protocol

Burns et al. [16] proposed the Meeting-Visit routing protocol (MV) that takes history information into consideration, such as frequency of meetings and their visits times at a certain location. The main forwarding selection criteria is based on the delivery probability $P_n^k(i)$, which calculates the delivery probability from current node k to the destination node i within n hops. Its formula is represented as,

$$P_n^k(i) = 1 - \prod_{j=1}^{N}(1 - m_{jk}P_{n-1}^j(i)),$$

where m_{jk} is the meeting probability of node j and node k during the last t round.

When node k meets another node, it exchanges a carried message list with its delivery probability and sorts the list based on the updated delivery probability. Node k removes its own message with a lower probability compared with the list obtained from its neighbors. After that, it selects the top n messages to forward to and request those un-stored messages from its neighbor as well.

3.3 Geographic Based Routing Protocols

3.3.1 Vector Routing Protocol

Kang and Kim proposed a vector routing in delay tolerant networks [17] using the nodes own location information to calculate their velocity and direction to destination, which reduces the overhead without decreasing the delivery success ratio. A mobile node uses its current vector V_{cur} and history movement pattern to predict a future vector. If two nodes are moving in the same direction, the replication of packets for both of nodes is not efficient. On the other hand, if they move in different directions, it is necessary to forward packets to both nodes. In vector routing, nodes moving in an orthogonal direction have a higher packet replication frequency compared to nodes moving in similar or opposite directions.

Furthermore, the velocity factor is also been taken into consideration. If two nodes move in the same direction, but the first node is faster than the second, packets may be forwarded to the faster node instead of replicating the slower node again. Simulations conducted from the ns-2 platform demonstrated that vector routing has a similar delivery success ratio compared with Epidemic routing, but with 28 % less traffic than random way-point model and 38 % less traffic than Manhattan mobility model.

3.3.2 Similarity Based Mobility Pattern Aware Routing Protocol

Yin et al. [18] proposed a similarity based mobility pattern aware routing in DNTs (SD-MPAR), which uses GPS to detect the location of nodes and calculates their mobility pattern, where the more identical the mobility pattern, the more stable their relative positions. The mobility pattern used in this paper is the intermediate node's (node r) speed direction to the destination node (node d), which is an angle (θ_{rd}).

Based on the nodes location information ((x_r, y_r) and (x_d, y_d)) and the weight controller (λ_i), a Similarity Degree ($f(r,d)$) is proposed to select the next forwarding node with the highest similarity degree in the source node's one hop neighbor area

$$f(r,d) = \lambda_1 \times \frac{1}{\theta_{rd}} + \lambda_2 \times \frac{1}{\sqrt{(x_r - x_d)^2 + (y_r - y_d)^2}}.$$

3.3.3   Location-Aware Routing Protocol

Tian and Li proposed [19] a location-aware routing protocol to study the node isolation issue in DTNs. In this routing protocol, analysis of both the location information and social network structure is conducted. When the current node and the destination node are within the same component and if the encountered node has a bigger metric than the current node, the packet is forwarded to this encountered node. Otherwise, the current node just wait next available nodes.

Furthermore, Tian and Li differentiated the operations of forwarding and replicating. In their paper, a node can forward a message to an encounter node once only and then it becomes forwarding inactive, even though it can replicate a message to other nodes without such constraint. Therefore, when the current node is in a different component to the destination node and if the current node has no previous contact with the encountered node, the packet is replicated to the encountered node. Otherwise, the packet is forwarded to the encountered node.

3.3.4   Location Aided Routing Protocol

Ko and Vaidya [20] proposed Location Aided Routing (LAR) to show how to use the location information to reduce routing overhead by limiting the search space for a certain route. The traditional routing discovery is based on the flooding process, where source nodes broadcast route request messages. When an inter-node receives this request message, it compares the message' destination with its own identifier. If they match, this message arrives at the final destination, otherwise the inter-node needs to broadcast this message again.

LAR uses location information to reduce the number of propagated nodes where they define two concepts: the expected zone and the request zone. The expected zone is a location where a certain node has visited before. With the expected zone in mind, the forwarding process can bypass lots of useless inter-nodes. If the source node has no idea where the expected zone is, the entire region then becomes the expected zone and the flooding process is deployed. The request zone defines a route requesting area (flooding area), which includes the expected zone and its surrounding zones. Therefore, the size of the request zone determines the routing performance and the network overhead. One potential method to determine the size of the request zone based LAR is to find the smallest rectangle that includes the current location and the expected zone. This expected zone is a circle area, where its center is the previous location of the destination node and its radius is the product of the time period and the node's movement speed. Therefore, the size of this request zone is subject to the average node's movement speed and the time elapsed since the last recorded known location.

3.3.5 Geography Aware Active Routing Protocol

Fan et al. [21, 22] proposed a geography aware active data dissemination protocol in human associated DTNs that analyzes mobile social networks from a geographic point of view. This paper uses the "community" concept from social science, which is defined as a clustering of users closely linked to each other [23] and extends it to "geo-community" by considering the relationship of nodes on geographic related social networks. The higher the centrality of a geo-community, the better capable it is of contacting mobile users in social networks. To quantify the dynamic density of a geo-community, the following geo-centrality technique is applied as,

$$C_i(t_i) = 1 - \frac{1}{N} \sum_{k=1}^{N} (1 - \phi_i^k)^{t_i},$$

where $C_i(t_i)$ indicates the average probability a certain user visits geo-community i at time t_i, and ϕ_i^k is used to measure the steady state probability distribution for node k in its ith geo-community. With this geo-community, the data forwarding process can be conducted by combing social and geographic information.

3.4 Social Concept Based Routing Protocols

3.4.1 Social Group Based Routing Protocol

Abdelkader et al. [24] proposed a Social Group based Routing (SGBR) protocol. SGBR is a multi-cast routing protocol that takes advantage of social grouping among networks to improve the delivery performance without increasing network traffic. If nodes belong to the same social group, they are expected to have the same social behaviours and meet each other more frequently. Therefore, every node in a social group may treat itself as a representative of its social group and tends to contact nodes in other social groups if it needs to send a packet outside of its own social group. Abedelkader et al. proved the delivery probability of replicating a message to other social groups is much higher than replicating the message within the same social group. It's forwarding strategy is to send packet to the node with high strength, which is measured by the degree of connectivity (τ_{ab}). τ_{ab} is used to quantify the meeting frequency between node a and node b. The following equation is the updating process of τ_{ab}, where α is the updating factor, $\alpha \in (0, 1]$, γ is the aging constant, $\gamma \in [0, 1]$, and k is the number of elapsed time since the last time two nodes met.

$$\tau_{ab} = (\tau_{ab})_{old} \gamma^k + (1 - (\tau_{ab})_{old} \gamma^k) \alpha$$

3.4.2   *Context-Aware Routing Protocol*

Context-Aware Routing (CAR) [25] utilizes contextual information to predict the probability of packet delivery, where the delivery probability is synthesized locally from contextual information. The contextual information may include mobility attributes, battery level, link loss ratio or the co-location attribute. In CAR, the current context information is useless, as it is changeable and dependants on the context of routing at the time. Instead, the routing decision of CAR is based on the prediction of future context attribute values.

To achieve the best attributes to represent the relevant aspects of the system for message delivery, the Weights Method [26] is applied to maximize the utility function for each of the attributes (U_i is a utility function of attribute x_i), as the following equation,

$$Maximise\, f(U(x_i)) = \sum_{i=1}^{n} w_i U_i(x_i),$$

where w_i is the significance weight to reflect the relative importance of each goal.

Once the maximized utility function is ready, it is applied by CAR to choose the node to deliver a message to its destination.

3.4.3   *Friendship Based Routing Protocol*

Bulut and Szymanski introduced a friendship based routing protocol in their work [27, 28]. This protocol measures three friendship behaviour features, frequency, longevity and regularity, from nodes' encounter history. The frequency means the average inter-connecting time and the regularity means the variance of the inter-connecting time.

For the directly connected nodes, a social pressure metric (SPM), equation 3.1, is used to measure a social pressure to motivate two nodes, node i and node j, to meet with each other, where $f(t)$ represents how much time is left before the next two nodes meet at time t. Friendship of two nodes is measured by the link quality which is the inverse of the SPM. The larger link quality value, the better the relationship is. Based on this link quality, the one hop friendship community can be defined as a set of nodes whose link qualities are bigger than a threshold.

$$SPM_{i,j} = \frac{\int_{t=0}^{T} f(t)dt}{T} \tag{3.1}$$

For indirectly connected nodes, node i contacts node k through node j, a relative social pressure metric (RSPM) equation is proposed, where the $t_{a,x}$ means there is number of x indirect contacts occurring at stage t_a. The link quality from node i to node k through node j is the inverse of $RSPM_{i,k|j}$,

$$RSPM_{i,k|j} = \frac{\sum_{x=1}^{n} \int_0^{t_{a,x}} (t_{b,x} + t_{a,x} - t)dt}{T}$$

$$= \frac{\sum_{x=1}^{n} (2t_{b,x}t_{a,x} + t_{t,x}^2)}{2T}.$$

Instead of directly detecting the SPM from two connected nodes, indirected nodes need to update PSPM whenever they meet. For example, to form a two hop friendship community, both values of link qualities need to be bigger than the threshold. The forwarding strategy of this routing protocol is to compare the friendship between current node and destination node, with the friendship between encountered node and destination node. As long as the encountered node has a stronger friendship and shares the same friendship community with destination node, packet is forwarded from current node to this encountered node. Otherwise, current node waits until the next available encountered node.

3.4.4 Practical Incentive Routing Protocol

Lu et al. [29] proposed a Practical incentive (Pi) routing protocol to address the selfishness problem in DTNs. Instead of avoiding selfish behaviours, this paper takes advantage of naturally selfish human behaviours to apply the layered coin model [30] to create a credit and reputation based incentive model. In this model, if an inter-node can successfully forward a packet to the destination, it is credited from the source node. Even if the destination node cannot receive the packet, the inter-node can still be credited by a trusted authority as long as it takes the forwarding process. The trusted authority obtains and analyses signatures from the last intermediate nodes, where a credit is calculated and charged from the source node's account to this forwarding node's account. At the meantime, an reputation value is calculated by the trusted authority and awarded to this forwarding node by using the incentive policy. The forwarding strategy of this routing protocol is based on location distances. When current node wants to forward a message to the destination, it waits a certain time for an available node to help it forwarding. If this available node's possible movement locations are closer to the destination compared with the current node, the message carried by current node is forwarded to this available node. Otherwise, this message is dropped if the waiting period exceeds the pre-defined threshold.

3.4.5 Social Network Oriented Duration Utility Based Distributed Multicopy Routing Protocol

Li and Shen [31] proposed a social network oriented duration utility based distributed multicopy routing protocol (SEDUM) for DTNs. The duration utility measures the coloration and familiar stranger attributes of nodes in mobile social network. This is the ratio of total contact duration between two nodes over a time period, where a higher duration utility indicates a higher message transmission throughput between two nodes. The duration utility for nodes i and j is

$$U_{(i,j)} = \frac{\sum_{k=1}^{fT} t_{(i,j)}(k)}{T},$$

where $t_{(i,j)}(k)$ indicates the encounter duration of the kth encounter, f is the contact frequency between nodes i and j, and fT indicates the total contacts during the time period T.

SEDUM has three phases: the replicating phase, the forwarding phase, and the clearing phase. The replicating phase is the initial broadcast process to send a replicated copy to different nodes. The forwarding phase forwards messages to another node with a higher duration utility. The clearing phase broadcasts discard message notifications once the destination node receives the message. The forwarding phase also utilizes the buffer management by introducing the core-replica concept. A core-replica has a higher priority stay in the buffer compared with non-core-replicas, with a core-replica having a higher delivery utility.

3.4.6 SocialCast Routing Protocol

Costa et al. [32] proposed a SocialCast routing protocol for publish-subscribe, where published messages are only sent to nodes with the same subscribed interests. The routing decision for SocialCast is based on the metrics of social interaction, such as connectivity changes. SocialCast has four phases: interest dissemination, carrier selection, message dissemination and message publishing. These are summarized as follows:

1. **Interest Dissemination:** nodes broadcast and exchange a summary vector, which includes a list of interests and associating utility values for its one hop neighbours.
2. **Carrier Selection:** The utility values of a certain interest for each of the one hop neighbours are compared, where the highest utility node is selected as the next carrier.
3. **Message Dissemination:** The content is re-evaluated based on the utilities and the message is forwarded to the best carrier.
4. **Message Publishing:** Once a message is published, it is stored into the local buffer.

3.4.7   Probabilistic Routing Protocol Using History of Encounters and Transitivity

PRoPHET (Probabilistic Routing Protocol using History of Encounters and Transitivity) was proposed by Lindgren et al. in [33]. In this work, it estimates a probabilistic metric, $P_{(a,b)}$, based on its previous probabilistic metric, $P_{(a,b)_{old}}$, for each node, a, to its every known destination b, as shown in equation 3.2. PRoPHET uses this metric to make a decision on how to select the next forwarding node in its local neighbours. When two nodes meet, a summary vector and a delivery predictability vector are exchanged, which contain the delivery predictability information for a destination known by the nodes.

$$P_{(a,b)} = P_{(a,b)_{old}} + (1 - P_{(a,b)_{old}}) \times P_{init} \qquad (3.2)$$

Furthermore, the delivery predictability of PRoPHET also has a transitive property that analyzes the encounter frequency among nodes a, b and c. If node a has a high encounter frequency with node b, and node b also has a high encounter frequency with node c, then node a is a potential node to forward messages to node c. The delivery probability between node a and node c is:

$$P_{(a,c)} = P_{(a,c)_{old}} + (1 - P_{(a,c)_{old}}) \times P_{(a,b)} \times P_{(b,c)} \times \beta \qquad (3.3)$$

For the above equations, both P_{init} and β are in a range of $[0, 1]$, where P_{init} is an initialization constant of $P_{(a,b)}$ and β is a scaling constant. Based on these probabilities, multiple destinations can be selected to forward messages to.

3.4.8   Simbet Routing Protocol

Daly and Haahr [34] proposed the combination of centrality, tie strengths, and tie predictors as highly useful in routing decision making, if the network exhibits a social structure. It uses SimBetTSUtil to combine these three factors, where the bridging capability of nodes is measured by the betweenness aspect of the SimBetTSUtil.

SimBetTSUtil calculates similarity and the betweenness centrality based on ego networks, and allows for distributed implementation. More importantly, egocentric calculations overcome the drawback of socio-centric networks to obtain local information, as socio-centric networks need to use global information. To summarize, compared with node m, $TSUtil_n$ is used to represent the tie strength utility, and $BetUtil_n$ is used to represent the betweenness utility, and $SimUtil_n$ represents the similarity utility of node n.

$$SimUtil_n(d) = \frac{Sim_n(d)}{Sim_n(d) + Sim_m(d)} \qquad (3.4)$$

$$BetUtil_n = \frac{Bet_n}{Bet_n + Bet_m} \qquad (3.5)$$

$$TSUtil_n(d) = \frac{TieStrength_n(d)}{TieStrength_n(d) + TieStrength_m(d)} \qquad (3.6)$$

The $SimBetTSUtil_n(d)$ is used to combine the above three factors and can be represented as,

$$SimBetTSUtil_n(d) = \sum_{u \in U} u_n(d),$$

where $U = \{TSUtil, SimUtil, BetUtil\}$. To select the delivery node from node n and node m, a comparison needs to be made about which one is larger from either $SimBetTSUtil_n(d)$ and $SimBetTSUtil_m(d)$.

3.4.9 Bubble Rap Routing Protocol

Hui et al. [35, 36] expanded the idea from the LABEL [37]. LABEL is the first routing protocol to demonstrate that incorporating a community affiliation label can improve forwarding performance. In this work, they proposed Bubble Rap routing protocol in a pocket switched network [38], a kind of mobile social network. It measures community affiliation labels with betweenness centrality to forward messages. A minimum of two centrality measures are calculated per node based on the node's global popularity in the whole network and its local popularity within its community or communities.

The Bubble algorithm can be simplified as follows: a message is always transferred to nodes with higher global rankings until the carrier encounters a node in the same community as the destination node. The local ranking system is then used and the message continues to bubble up through the local ranking tree until the destination is reached or the message expires. Nodes in this model only need to compare rankings with encountered nodes instead of requiring every node to know the ranking of others.

Furthermore, the authors introduced the unit-time degree to measure the contact frequency based on different time periods, including both single window and cumulative window. The single window is used to compare how many unique nodes have been met in the previous unit-time slot. The cumulative window is used to calculate the average degree for every single window.

3.5 Time Related Routing Protocols

3.5.1 Transient Social Contact Pattern

Gao et al. [39, 40] proposed the use of transient social contact patterns to improve the data forwarding performance. This model has three parts: transient contact distribution, transient connectivity, and transient·community structure. Transient contact distribution differentiates contact distributions for different time periods, and is the main trade off compared to cumulative contact distribution. For example, classmates have a high contact frequency during the daytime, instead of night-time.

Transient connectivity identifies the connection of nodes during a certain time period that can be used to form a transient connected subnet (TCS). For example, students can form a TCS with their classmates during the day in class, while form another TCS with their parents during the night at home. A transient community structure is used to identify how many communities a node belong to during a certain period. By using this transient community structure, the centrality of nodes can be quantified more accurately.

The main benefit of using this transient social contact is to introduce a time pattern into a social community so the importance of each social attribute can be quantified better.

3.5.2 Time Evolving Topology Control

Huang et al. [41, 42] modelled a time evolving network to include both geographic and time information in a space time graph. The topology of delay tolerant networks are changeable due to their natural characteristics, where mobile nodes always move around. These changes are time dependent, and usually modelled as a connected graph with stable end to end paths.

Two greedy-based methods, one based on the least cost path and another based on a least density bunch, were proposed to reduce the total cost of network topology, where a set of edges were repeatedly added into the topology to connect one or more pairs of nodes in their space-time graph. The main trade off for the second method compared with the first is to repeatedly find good bunches instead of repeatedly finding least costy paths.

References

1. Perkins, C.E., Bhagwat, P.: Highly dynamic destination-sequenced distance-vector routing (dsdv) for mobile computers. In: Proceedings of the Conference on Communications Architectures, Protocols and Applications (SIGCOMM '94), London, pp. 234–244. ACM, New York (1994)

2. Dhurandher, S., Obaidat, M., Gupta, M.: A reactive optimized link state routing protocol for mobile ad hoc networks. In: Proceedings of IEEE ICECS, Athens, pp. 367–370 (2010)
3. Perkins, C., Royer, E.: Ad-hoc on-demand distance vector routing. In: Proceedings of IEEE WMCSA, New Orleans, pp. 90–100, Feb 1999.
4. Wu, J.: An extended dynamic source routing scheme in ad hoc wireless networks. In: Proceedings of Annual Hawaii International Conference on System Sciences, Big Island, pp. 3832–3838 (2002)
5. Lindgren, A., Doria, A., Schelén, O.: Probabilistic routing in intermittently connected networks. SIGMOBILE Mob. Comput. Commun. Rev. 7, 19–20 (2003)
6. Vogels, W., van Renesse, R., Birman, K.: The power of epidemics: robust communication for large-scale distributed systems. SIGCOMM Comput. Commun. Rev. 33, 131–135 (2003)
7. Harras, K.A., Almeroth, K.C., Belding-Royer, E.M.: Delay tolerant mobile networks (DTMNs): controlled flooding in sparse mobile networks. In: Proceedings of IFIP-TC6 International Conference on Networking Technologies, Services, and Protocols (NETWORKING'05), Waterloo, pp. 1180–1192, Springer-Verlag, Berlin/Heidelberg (2005)
8. Zhu, H., Chang, S., Li, M., Naik, K., Shen, X.: Exploiting temporal dependency for opportunistic forwarding in urban vehicular networks. In: Proceedings IEEE INFOCOM, Shanghai, pp. 2192–2200 (2011)
9. Davis, J.A., Fagg, A.H., Levine, B.N.: Wearable computers as packet transport mechanisms in highly-partitioned ad-hoc networks. In: Proceedings of IEEE ISWC, Zurich (2001)
10. Mundur, P., Seligman, M., Lee, G.: Epidemic routing with immunity in delay tolerant networks. In: Proceedings of IEEE MILCOM, San Diego, pp. 1–7 (2008)
11. Tower, J.P., Little, T.D.C.: A proposed scheme for epidemic routing with active curing for opportunistic networks. In: Proceedings of IEEE AINAW, Gino-wan, pp. 1696–1701 (2008)
12. Spyropoulos, T., Psounis, K., Raghavendra, C.S.: Spray and wait: an efficient routing scheme for intermittently connected mobile networks. In: Proceedings of ACM SIGCOMM workshop on Delay-tolerant networking (WDTN '05), pp. 252–259. ACM, New York, Philadelphia, PA (2005)
13. Spyropoulos, T., Psounis, K., Raghavendra, C.S.: Single-copy routing in intermittently connected mobile networks. In: Proceedings of the First Annual IEEE Communications Society Conference on Sensor and Ad Hoc Communications and Networks (SECON 2004), Santa Clara, pp. 235–244, IEEE Press, (2004)
14. Paule, P., Schneider, C.: Computer proofs of a new family of harmonic number identities. Adv. Appl. Math. 31(2), 359–378 (2003). Preliminary version online
15. Ramanathan, R., Hansen, R., Basu, P., Rosales-Hain, R., Krishnan, R.: Prioritized epidemic routing for opportunistic networks. In: Proceedings of ACM MobiOpp, San Juan, pp. 62–66. ACM, New York (2007)
16. Burns, B., Brock, O., Levine, B.: MORA routing and capacity building in disruption tolerant networks. In: Proceedings of the 24th Annual Joint Conference of the IEEE Computer and Communications Societies (INFOCOM 2005), Miami, vol. 1, pp. 398–408. IEEE (2005)
17. Kang, H., Kim, D.: Vector routing for delay tolerant networks. In: Proceedings of IEEE VTC, Marina Bay, Singapore pp. 1–5 (2008)
18. Yin, L., Cao, Y., He, W.: Similarity degree-based mobility pattern aware routing in DTNs. In: Proceedings of the 2009 International Symposium on Intelligent Ubiquitous Computing and Education (IUCE '09), Chengdu,, pp. 345–348. IEEE Computer Society, Washington, DC (2009)
19. Tian, Y., Li, J.: Location-aware routing for delay tolerant networks. In: Proccedings of ICST CHINACOM, pp. 1–5, Beijing, China (2010)
20. Ko, Y.-B., Vaidya, N.H.: Location-aided routing (LAR) in mobile ad hoc networks. Wirel. Netw. 6, 307–321 (2000)
21. Fan, J., Du, Y., Gao, W., Chen, J., Sun, Y.: Geography-aware active data dissemination in mobile social networks. In: Proceedings of IEEE MASS, San Francisco, pp. 109–118 (2010)

22. Fan, J., Chen, J., Du, Y., Gao, W., Wu, J., Sun, Y.: Geocommunity-based broadcasting for data dissemination in mobile social networks. IEEE Trans. Parallel Distrib. Syst. **24**, 734–743 (2013)
23. Yang, Z., Tang, J., Li, J., Yang, W.: Social community analysis via a factor graph model. IEEE Intell. Syst. **26**(3), 58–65 (2011)
24. Abdelkader, T., Naik, K., Nayak, A., Goel, N., Srivastava, V.: SGBR: A routing protocol for delay tolerant networks using social grouping. IEEE Trans. Parallel Distrib. Syst. **24**(12), 2472–2481 (2013)
25. Musolesi, M., Mascolo, C.: CAR: Context-aware adaptive routing for delay-tolerant mobile networks. IEEE Trans. Mobile Comput. **8**, 246–260 (2009)
26. Jong, N.K., Stone, P.: Keeney,r.l. raiffa,h.decisions with multiple objectives: Preferences and value tradeoffs. In: Proceedings of ICML-06 Workshop on Kernel Methods in Reinforcement Learning. Wiley, Pittsburgh, Pennsylvania (1976)
27. Bulut, E., Szymanski, B.K.: Exploiting friendship relations for efficient routing in mobile social networks. IEEE Trans. Parallel Distrib. Syst. **23**, 2254–2265 (2012)
28. Bulut, E., Szymanski, B.: Friendship based routing in delay tolerant mobile social networks. In: Proceedings of IEEE GLOBECOM, Miami, pp. 1–5 (2010)
29. Lu, R., Lin, X., Zhu, H., Shen, X., Preiss, B.: Pi: a practical incentive protocol for delay tolerant networks. IEEE Trans. Wirel. Commun. **9**(4), 1483–1493 (2010)
30. Zhu, H., Lin, X., Lu, R., Ho, P.-H., Shen, X.: SLAB: A secure localized authentication and billing scheme for wireless mesh networks. IEEE Trans. Wirel. Commun. **7**(10), 3858–3868 (2008)
31. Li, Z., Shen, H.: SEDUM: Exploiting social networks in utility-based distributed routing for DTNs. IEEE Trans. Comput. **62**(1), 83–97 (2013)
32. Costa, P., Mascolo, C., Musolesi, M., Picco, G.: Socially-aware routing for publish-subscribe in delay-tolerant mobile ad hoc networks. IEEE J. Sel. Areas Commun. **26**(5), 748–760 (2008)
33. Lindgren, A., Doria, A., Schelén, O.: Probabilistic routing in intermittently connected networks. SIGMOBILE Mobibile Comput. Commun. Rev. **7**, 19–20 (2003)
34. Daly, E., Haahr, M.: Social network analysis for information flow in disconnected delay-tolerant MANETs. IEEE Trans. Mobile Comput. **8**, 606–621 (2009)
35. Hui, P., Crowcroft, J., Yoneki, E.: BUBBLE Rap: Social-based forwarding in delay-tolerant networks. IEEE Trans. Mobile Comput. **10**, 1576–1589 (2011)
36. Hui, P., Crowcroft, J., Yoneki, E.: Bubble rap: social-based forwarding in delay tolerant networks. In Proceedings of ACM MobiHoc, Hong Kong, pp. 241–250 (2008)
37. Hui, P., Crowcroft, J.: How small labels create big improvements. In Proceedings of IEEE PerCom, White Plains, pp. 65–70 (2007)
38. Hui, P., Chaintreau, A., Scott, J., Gass, R., Crowcroft, J., Diot, C.: Pocket switched networks and human mobility in conference environments. In: Proceedings of ACM SIGCOMM Workshop on Delay-Tolerant Networking, pp. 244–251, ACM, New York, Philadelphia, PA (2005)
39. Gao, W., Cao, G.: On exploiting transient contact patterns for data forwarding in delay tolerant networks. In: Prceedings of IEEE ICNP, Kyoto, pp. 193–202 (2010)
40. Gao, W., Cao, G., La Porta, T., Han, J.: On exploiting transient social contact patterns for data forwarding in delay-tolerant networks. IEEE Trans. Mobile Comput. **12**, 151–165 (2013)
41. Huang, M., Chen, S., Zhu, Y., Xu, B., Wang, Y.: Topology control for time-evolving and predictable delay-tolerant networks. In: 2011 IEEE Eighth International Conference on Mobile Ad-Hoc and Sensor Systems, Valencia, pp. 82–91. IEEE, Los Alamitos (2011)
42. Huang, M., Chen, S., Zhu, Y., Wang, Y.: Topology control for time-evolving and predictable delay-tolerant networks. IEEE Trans. Comput. (to appear)

Chapter 4
Social Characteristics Based Multiple Dimensional Routing Protocol in Human Associated Delay Tolerant Networks

This chapter introduces how to use multiple dimensional social attributes to improve data forwarding performance in human associated delay tolerant networks. As mobile nodes in these human associated DTNs are carried by humans, data forwarding processes are determined by human social behaviors [1]. If a node has two or more social attributes, it can be treated as a social hub (popular nodes with many social attributes), where characteristics with other nodes are shared in the overlapping area of the attributes.

In Fig. 4.1, for example, a simple hierarchical work structure in a school is shown. Individuals are sorted by their affiliations within the faculty. If student A wants to contact the Head of School, as they are not connected directly, student A can either go through lecturer B and unit chair D or tutor C and unit chair E to the Head of School. Either path needs two inter-nodes. On the other hand, as shown in Fig. 4.2, people are directly linked by their social habits, where student A and person X are football club members, and Head of School and person X both swim in the same club. In this case, if student A wants to send a message to the Head of School, he can do so through person X, which is only need one hop compared to the previous example. This example demonstrates the importance of social characteristics if mobile devices are carried by humans.

The social characteristics based multiple dimensional model includes two parts: the social distance calculation and the multi-cast forwarding. The first part introduces the method to combine different social attributes from different dimensions that includes two sub-models. One combination is the three most useful dimensions into a single coordinate system, called the Triangle (3-Dimensional) coordinate system [2]. The second sub-model is the M-Dimension model [3] which is a general model to include as many social dimensions as possible. The multi-cast selection and forwarding process uses a social distance calculation function to compare and generate all available routes in its one hop neighbor sets and perform the multi-cast process. Briefly, the multiple dimension routing scheme can be described as a re-modeling of the network, assigning a random coordinate to each social dimension.

© The Author(s) 2015
L. Gao et al., *Delay Tolerant Networks*, SpringerBriefs in Computer Science,
DOI 10.1007/978-3-319-18108-0_4

Fig. 4.1 Example of a hierarchical structure within a school network

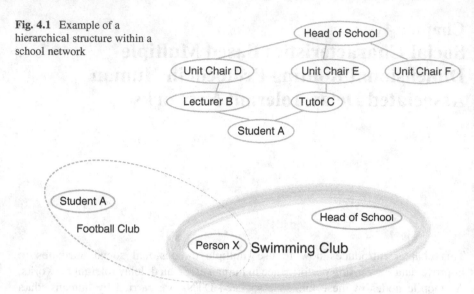

Fig. 4.2 Connections from a social habit point of view based on Fig. 4.1

Each mobile node is assigned an identification, vector formed by attributes in different social dimensions. The social distance between two nodes is computed by summing of the different attributes in each dimension.

4.1 Background

In this background section, social concepts used for this chapter are introduced firstly, which are homophily, small world theory and key paths. After that, a real trace file in DTNs is analysed to study the impact and importance of social characteristics in DTNs.

4.1.1 Social Concepts

4.1.1.1 Homophily

Similar characteristics between individuals generally result in a bond between two people. The saying "birds of a feather flock together" suggests [4] people with different characteristics, such as gender, age or educational background tend to have very different qualities. Alternatively, people with similar interests, backgrounds or beliefs tend to form stronger relationships than those with dissimilar ones [5].

Individuals with less in common are less likely to develop or maintain a strong tie, and therefore a pathway of communication between them will not be as strong. This phenomenon, known as homophily [4], suggests contact between similar people occurs at a higher rate than dissimilar people and their social characteristics are translated into distance. For example, people with a common interest may belong to the same social club and physical space at the same time.

The structure of a social network is greatly influenced by this principle. Individuals form groups or cliques in a given social network largely depending on their shared attributes. These attributes include similar interests, backgrounds, socioeconomic status, family ties, generation, age, affiliation, career, education, etc. Because such similarities connect (and disconnect) nodes in a powerful way, it dictates the flow of information in a given social network. Therefore efficient message pathways with fewer hops are inevitably created and reused if deemed to be dependable over time.

4.1.1.2 Small World Theory

Small world theory [6], which is often referred to as the "six degrees of separation", suggests every individual can be connected to another individual through a chain of individuals who know each other. Essentially, each person is connected to every other through a number of connections or "hops". An experiment conducted by Milgram [6] proved the "six degrees of separation" theory. In this experiment, several letters were sent to one stockbroker in Boston. Milgram distributed these letters to a random selection of people in Nebraska and Kansas, with instructions for the letters to be sent to the stockbroker by passing them from person to person. In addition, the letters could only be sent to someone whom the current holder knew on a first-name basis. The result of Migram's experiment showed that only 25 % of missives reached the final destination.

This experiment also demonstrated that our closest connections are stronger than our furthest connections. A direct connection in a social network means that two individuals know each other directly. For example, person A and person B are directly connected. If A wishes to send a message to B, the most efficient way to do so is to use the existing direct path between them. Through A, B can be made aware of all of A's direct connections. If B wishes to send a message to any of them, she would most likely do so using her direct path to A, and A's direct path to the desired node.

4.1.2 Case Study

The Infocom 06 dataset is one of benchmark datasets used in delay tolerant networks, which is part of the Huggle Project [7]. This dataset is a real social trace file recorded during 3 days of the Infocom 2006 conference, held at Princesa Sofia

Table 4.1 Summarization of the Infocom 06 dataset

Device	iMote
Duration in days	3
No. of nodes	98
Total edges	129,191
Maximum geodesic distance	2
Average geodesic distance	1.002724
Graph density	0.984420643

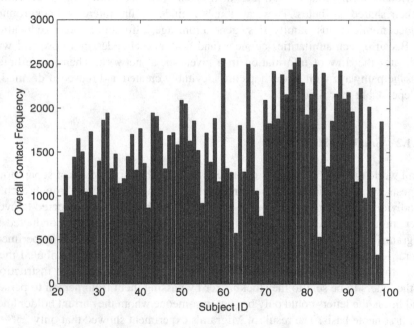

Fig. 4.3 The overall contact frequency of the Infocom 06 dataset

Gran hotel of Barcelona [8]. The main characteristics of this dataset are summarized in Table 4.1 and its overall contact frequency is plotted in Fig. 4.3. Selected people were asked to carry an Imote device, a portable Bluetooth transmitter, all the time no matter where they went.

The original trace file contained 98 nodes, of which 20 were stationery. Every user was asked to complete a questionnaire with personal information, such as nationality, first language, and professional affiliation. This questionnaire provided those social attributes needed in the experiment.

As these Bluetooth devices were carried by humans, their data forwarding process was supposed to have some social characteristics. To verify this dataset's social behaviors, 10 subjects were randomly selected from the 78 participants and each subject's name (Subject ID), nationality, study (the school where the participants completed her or his study), languages (at most three languages were listed), affiliation, position (such as student, researcher or professor), city (city and

Table 4.2 Ten subjects from Infocom 06 dataset

Subject ID	Nationality	Studies	1st language	2nd language	3rd language
25	4	18	5	1	14
34	1	35	1	2	n/a
39	3	3	5	n/a	n/a
46	26	3	7	1	5
56	2	2	3	1	4
61	n/a	n/a	n/a	n/a	n/a
66	16	36	14	1	n/a
72	14	19	22	1	15
76	16	36	14	1	n/a
87	14	19	15	1	n/a

Table 4.3 Ten subjects from Infocom 06 dataset (continued)

Subject ID	Affiliation	Position	City	Country	Stay
25	13	1	5	5	7
34	1	1	1	1	1
39	3	3	3	3	2
46	3	3	3	3	2
56	2	2	2	2	2
61	n/a	n/a	n/a	n/a	n/a
66	32	3	24	15	15
72	23	3	23	14	4
76	24	3	24	15	n/a
87	23	3	23	14	4

the country were used to identify the participant's residence status), country and stay (indicates the hotel where the participant was staying during the conference) are collected, as shown in Tables 4.2 and 4.3. In these tables, if two subjects have the same value for a certain attribute, that means these two subjects have the same attribute. For these ten subjects, the contact frequency amongst each other during the 3 day conference is listed in Table 4.4.

From Table 4.4, the number of contact amongst each other for most participants is less than 50. However, some pairs, such as Subject 39 and Subject 46, Subject 66 and Subject 76, and Subject 72 and Subject 87 contacted more than 100 times. The reason behind this phenomenon is due to their social attributes. For subjects 39 and 46, the attributes they shared were language (Language 5), affiliation, position, city, country and stay, meaning they most likely come from the same research lab and speak the same language. For subjects 66 and 76, the attributes they shared were nationality, study, language, position, city and country, meaning they most likely come from the same city and have the same nationality. For subjects 72 and 87, they shared all attributes listed. These results demonstrate the Infocom 06 dataset has strong social behaviours and these social behaviours have strong impact on subjects' contacts.

Table 4.4 Contact frequency among the ten subjects in the Infocom 06 dataset

Subject ID	25	34	39	46	56	61	66	72	76	87
25		4	18	23	50	13	22	30	15	20
34	1		8	9	2	17	35	17	46	12
39	18	6		128	22	11	18	31	18	25
46	18	4	125		19	9	8	25	8	14
56	41	2	10	18		41	22	18	18	14
61	7	15	5	2	46		16	9	32	7
66	13	38	14	10	14	20		47	117	41
72	28	18	29	27	23	15	45		51	116
76	11	38	15	6	12	28	148	53		41
87	33	10	26	20	23	21	46	105	36	

It is more likely an interaction will take place between two people who speak the same language, have the same field of interest or come from the same school, compared with individuals who share less personal information, such as belonging to the same professional association like IEEE. As the homophily phenomenon [4] demonstrated, different social attributes have different importance. Furthermore, even for the same attribute, different values represent different social performance. For example, in Fig. 4.4, Language Attribute only has very limited "key" factors, such as Language 1 and 5, where it follows the power law distribution [9], as shown in Fig. 4.5 and its associated subject mapping is listed in Table 4.5. Equation 4.1 shows the power law distribution function and its parameters are listed in Table 4.6.

$$f(x) = ax^k + \varepsilon \tag{4.1}$$

More importantly, social activities in a given area show a higher level of interaction among similar people. As a strong example, the language attribute in the Infocom 06 dataset demonstrates that while English was the main language used at this conference, it was easy to distinguish those who spoke Chinese (Language 5) as they tended to interact more often with each other. On another hand, if English was not involved, most people tended to have over 26 % of their total interactions with people who spoke the same language, as shown in Fig. 4.6. As every participant spoke English (Language 1), it means English is useless to compare the social similarity for the Language attribute.

4.2 3-Dimensional Coordinate Model

4.2.1 Triangle Coordination System

The 3-Dimensional Coordinate Model puts the three most useful social attributes together by calculating the pair-wise correlations to obtain the relationships among

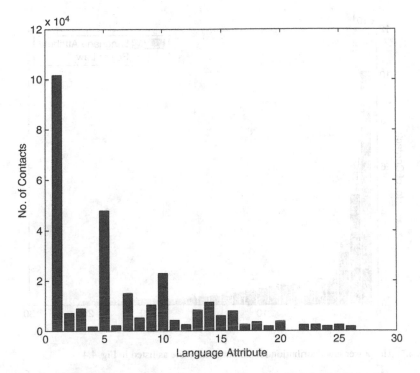

Fig. 4.4 Overall contact frequency of the language attribute for the Infocom 06 dataset

the three attributes. In general, for a 3-dimensional coordinate system, the three data series in vectors can be defined as X, Y, and Z, as shown in Fig. 4.7. To form a triangle, an adjacent side or a hypotenuse needs to be determined. As the side of the hypotenuse is greater than the adjacent side, it can be distinguished by using the following equations to adjust the increments in X and Y, where the T is represented as matrices/vectors transpose.

- $(X^T X)^{-1} X^T Y > 1$ means when X increases by 1 unit, Y increases by more than 1 unit, which indicates that X is the adjacent side;
- $(X^T X)^{-1} X^T Y < 1$ means when X increases by 1 unit, Y increases by more than 1 unit, which indicates that X is the hypotenuse;
- $(X^T X)^{-1} X^T Y = 1$ means the angle between X and Y is already a right angle and there is no need for scaling.

To form a triangle, Z needs to satisfy the following two conditions:

$$(X^T X)^{-1} X^T Z < 1 (Y^T Y)^{-1} Y^T Z < 1$$

Therefore, the estimated angle between Y and X is

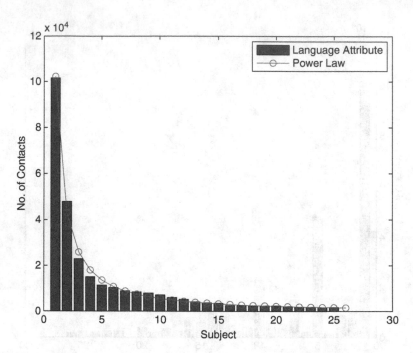

Fig. 4.5 The power law distribution for language attribute as listed in Fig. 4.4

Table 4.5 Language attribute mapping table for Fig. 4.5

Subject	Language attribute	Subject	Language attribute
1	Language 1	14	Language 20
2	Language 5	15	Language 18
3	Language 10	16	Language 17
4	Language 7	17	Language 12
5	Language 14	18	Language 23
6	Language 9	19	Language 22
7	Language 3	20	Language 25
8	Language 13	21	Language 6
9	Language 16	22	Language 24
10	Language 2	23	Language 19
11	Language 15	24	Language 26
12	Language 8	25	Language 4
13	Language 11	26	Language 21

Table 4.6 Power law distribution parameter of Eq. 4.1 for Fig. 4.5

Parameter	Value
a	1.025e+005
k	−1.246
ε	−238.1

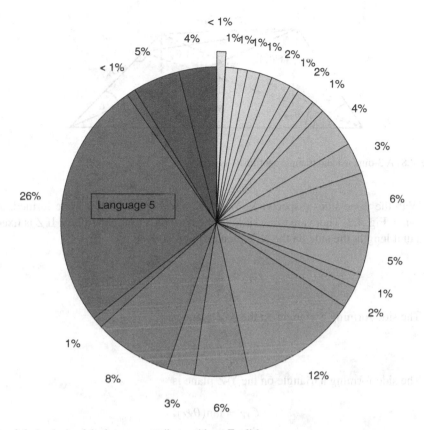

Fig. 4.6 Analysis of the language attribute without English

Fig. 4.7 A 3-dimensional
data series

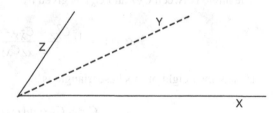

$$\theta_{YX} = \cos^{-1}[(Y^T Y)^{-1} Y^T X],$$

and the estimated angle between Y and Z is

$$\theta_{YZ} = \cos^{-1}[(Y^T Y)^{-1} Y^T Z],$$

and the estimated angle between X and Z is

$$\theta_{XZ} = \cos^{-1}[(X^T X)^{-1} X^T Z].$$

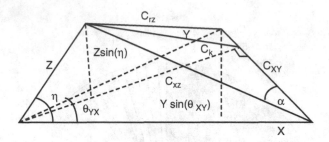

Fig. 4.8 A 3-dimensional triangle coordinate system

With the three sides (aspects) X, Y and Z, a triangular body can be formed, as shown in Fig. 4.8. There are three sides of the base on a triangular body. If Z is fixed as a unit length, the side on the X-Y plane can be derived by

$$C_{XY} = \frac{\tan(\theta_{YX})}{\cos(\theta_{YZ})}.$$

The side forming a triangle on the X-Z plane is

$$C_{XZ} = \tan(\theta_{XZ}).$$

The side forming a triangle on the Y-Z plane is

$$C_{YZ} = \tan(\theta_{YZ}).$$

The angle between C_{XZ} and C_{XY} is given by

$$\alpha = \cos^{-1}\left(\frac{C_{XZ}^2 + C_{XY}^2 - C_{YZ}^2}{2C_{XZ}C_{XY}}\right).$$

Hence, the height of this base triangle is

$$C_K = C_{XZ}\sin(\alpha).$$

With C_K, the angle between Z and X-Y plane can be calculated as

$$\eta = \tan^{-1}(C_K).$$

Therefore, X is kept as the x-axis. To rotate Z to the z-axis, which is perpendicular to both the x and the y axis, it needs to be multiplied by $\sin(\eta)$, that is, all data pointing to Z is scaled by a factor of $\sin(\eta)$. To rotate Y to the y-axis, which is perpendicular to both the x and the z axis, Y needs to be multiplied by $\sin(\theta_{YX})$, that is, every data point in Y will be scaled by a factor of $\sin(\theta_{YX})$.

4.2.2 Social Distance Calculation

All individuals have their own position in this xyz 3-dimensional coordinate. For a particular individual i, its position vector in the xyz coordinate will be

$$V(i) = (X_i, Y_i \sin(\theta_{YX}), Z_i \sin(\eta)).$$

Hence, the social distance between node i and node j can be defined as Function 4.2

$$dist(i,j) = \sqrt{(X_i - X_j)^2 + (Y_i - Y_j)^2 \sin^2(\theta_{YX}) + (Z_i - Z_j)^2 \sin^2(\eta)} \qquad (4.2)$$

4.3 M-Dimension: Multi-dimensional Modeling

In M-Dimension, the socializing network structure is built on top of the geographic topology. Nodes are grouped into different clusters according to their social characteristics, such as occupation and interests. The foundation of M-Dimension is to model a network from different social dimensions. Attributes in social dimensions are taken from social concepts covered in Sect. 4.1.1. However, some homophily characteristics are more effective than others. For example, age is less effective than family in linking nodes since the possible values of age are numerous and therefore, can often fail in linking isolated groups of nodes as effectively as family links. Individuals are more likely to have a strong relationship tie with someone in the same family instead of someone who only shares the same age. Thus a high weight value can be assigned to the family dimension, where the weight factor is used to represent the importance of each dimension.

4.3.1 Unification Process and Weight Factor

Each node in this model can be identified as a H-dimensional coordinate vector, $v(n)$ with the h-tuple,

$$v(n) = < v_1, v_2, \ldots, v_h >, \qquad (4.3)$$

where v_i is the attribute value of node n in the ith dimension.

To obtain local information, M-Dimension uses Probe Messages to exchange local information of nodes and store this information within the local memories of nodes. Each Probe Message contains local information of the source node and a sequence number used by the receiver to identify the duplicated Probe

Messages. Nodes exchange their routing information with their local neighbours at a higher frequency. The shorter the distance, the more Probe Messages generated. This approach simulates the social phenomenon, where social nodes are likely to maintain up-to-date information of nodes that share similar social interests.

Following the definition from Watts et al. in [10], the distance of nodes on the same level is equal to 0. As the example in Fig. 4.1 shows, each level between layers delimits a distance equal to 1, meaning the distance between the Head of School and Unit Chair is 1, or the Head of School and student A is 3. To clarify this concept, when applied to a real social environment, if a Head of School has to communicate with the student A, the distance is used to define how complex the path is. As in real life, the Head of School has little contact with students, therefore the probability of them meeting with each other is relative low. Instead, they can use inter-nodes in other social communities to form a path reflecting the real life behaviour in a social network, as shown in Fig. 4.2.

When combining and calculating social distances among different dimensions, it can not directly use the function proposed by Watts et al. in [10]. Instead, all dimensions should be unified first and different weights should be applied to quantify the most important dimension.

The unification process is applied to calculate the significance of the individual dimension compared with the whole dimension, including all nodes. The individual dimension has its own unique identification, and representative values may vary for each dimension. After the unification process, each individual should be represented by a percentage vector instead of the original vector as listed in Eq. 4.3. For example, assume network A has n nodes and m dimensions. This network can be represented as:

$$A = \begin{bmatrix} A_1^1 & A_2^1 & \cdots & A_m^1 \\ A_1^2 & A_2^2 & \cdots & A_m^2 \\ A_1^3 & A_2^3 & \cdots & A_m^3 \\ \vdots & \vdots & \vdots & \vdots \\ A_1^n & A_2^n & \cdots & A_m^n \end{bmatrix}$$

For a certain node in network A, the percentage of each individual dimension taken in the whole dimension is calculated as:

$$A' = \begin{bmatrix} \dfrac{A_1^1}{\sum_{i=1}^n A_1^i} & \dfrac{A_2^1}{\sum_{i=1}^n A_2^i} & \cdots & \dfrac{A_m^1}{\sum_{i=1}^n A_m^i} \\ \dfrac{A_1^2}{\sum_{i=1}^n A_1^i} & \dfrac{A_2^2}{\sum_{i=1}^n A_2^i} & \cdots & \dfrac{A_m^2}{\sum_{i=1}^n A_m^i} \\ \dfrac{A_1^3}{\sum_{i=1}^n A_1^i} & \dfrac{A_2^3}{\sum_{i=1}^n A_2^i} & \cdots & \dfrac{A_m^3}{\sum_{i=1}^n A_m^i} \\ \vdots & \vdots & \vdots & \vdots \\ \dfrac{A_1^n}{\sum_{i=1}^n A_1^i} & \dfrac{A_2^n}{\sum_{i=1}^n A_2^i} & \cdots & \dfrac{A_m^n}{\sum_{i=1}^n A_m^i} \end{bmatrix}$$

As mentioned at the beginning of this section, even though two dimensions both have homophilic characteristics, one may be more important than the other. In order to measure these differences, the weighting process is applied to calculate weight factor, w, which identifies how important a certain dimension is as opposed to another dimension for a certain node. Assume a node has h dimensions in total and the overall weight factors need to satisfy the following rule:

$$\sum_{u=1}^{h} w_u = 1$$

For a certain node j, the weight factor in its q dimension is:

$$w_q = \frac{\frac{A_q^j}{\sum_{i=1}^{n} A_q^i}}{\frac{A_1^j}{\sum_{i=1}^{n} A_1^i} + \frac{A_2^j}{\sum_{i=1}^{n} A_2^i} + \cdots + \frac{A_m^j}{\sum_{i=1}^{n} A_m^i}}$$

$$= \frac{A_q^j}{\sum_{i=1}^{n} A_q^i \times \sum_{v=1}^{m} \frac{A_v^j}{\sum_{i=1}^{n} A_v^i}}$$

As long as the unification process is completed and the weight factors are collected, the overall social distance, with multiple dimensions involved, between two nodes can be generated.

4.3.2 Social Distance Calculation

The social distance between two nodes is based on unified values and weight factors to calculate the overall distance from candidature set to destination, where the candidature set is the one hop neighbor source set. For example, in Fig. 4.9, the candidature set for node S includes nodes $M_1, M_2, \cdots, M_h$. Therefore vectors for nodes S's destination and candidature set are:

$$v(D) = <D_1, D_2, \cdots, D_h>$$
$$v(M_1) = <M_1^1, M_2^1, \cdots, M_h^1>$$
$$v(M_2) = <M_1^2, M_2^2, \cdots, M_h^2>$$
$$\cdots\cdots$$
$$v(M_k) = <M_1^k, M_2^k, \cdots, M_h^k>$$

As the target of this model is to efficient route a packet to the destination, the destination's weight factors are used as domain factors. After the unification process, the unified destination vector $v(D)'$is:

$$v(D)' = < \frac{D_1}{\sum_{i=1}^{k} M_1^i + D_1}, \frac{D_2}{\sum_{i=1}^{k} M_2^i + D_2},$$

$$\ldots, \frac{D^h}{\sum_{i=1}^{k} M_h^i + D_h} >,$$

and the weight factor for the destination in dimension u ($u \leq h$) is represented as:

$$w_u = \frac{\frac{D_u}{\sum_{i=1}^{k} M_u^i + D_u}}{\sum_{r=1}^{h} \frac{D_r}{\sum_{i=1}^{k} M_r^i + D_r}} \qquad (4.4)$$

With the above information, the social distance between a candidature node M_i and destination node D is calculated by the $dist(M_i, D)$ function, as:

$$dist(M_i, D) = \sum_{u=1}^{h} w_u \times M_u^i \times D_u \qquad (4.5)$$

4.4   Multi-cast Selection and Forwarding

In traditional forwarding processes, a simple task is to choose where to forward to. This requires choosing the source node's neighbor with the shortest path to destination. In normal circumstances, messages are only sent to a single node, as the reliability of these paths is relatively high. However as the topology is always changing in DTNs, when a node receives a message, it may not be able to find an available path to the destination at that particular time period, and therefore the message needs to be stored for a while until another node arrives to its neighbor area. This means, the decision making process for DTNs can only be made in that particular situation. Hence, it requires multi-cast to increase the probability a message will be delivered to its destination.

In human associated DTNs, the multi-cast selection process is based on the Social Distance Function (Function 4.5) to select suitable inter-nodes from a candidature set. These selected inter-nodes are used to store and forward packets.

The details of the selection and forwarding process is described in Algorithm 1, where the source, node S, generates all available social distances from its candidature set to their destinations. Based on the number of nodes in candidature sets and the network change speed, an estimated delivery percentage is given to direct how many inter-nodes need to be selected and forwarded. Once inter-nodes are selected, node S multi-casts packets to them. When an inter-node receives a packet, it checks the packet's time stamp and the TTL to decide whether or not to multi-cast this packet again.

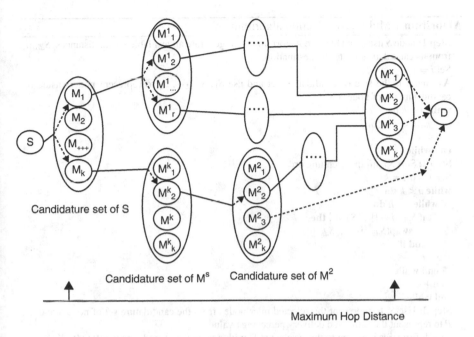

Fig. 4.9 The multi-cast model

4.5 Summary

This chapter identifies the importance of social characteristics in human associated DTNs and analyzes how to combine them by weighting the social attributes.

Two social characteristics based multi-dimensional routing protocols in human associated DTNs are proposed: the Triangle Coordinate model and M-Dimension model. Both of these models take different social characteristics into account from multiple dimensions and demonstrates how important each dimension is. For the Triangle Coordinate model, the three most useful social attributes are taken and put together into a triangle coordinate system to generate an overall social distance. For the M-Dimension model, the weighting process and distance function are deployed to calculate social distances between two nodes in each dimension with associated weight factors. Based on these distances, multi-cast routing can be performed. This notes the number of available candidatures and network change speed.

Algorithm 1 Multi-cast selection algorithm

Step 1: node S uses the Distance Algorithm 4.5 to generate all available social distances, $S_{dist}[i]$, from its candidature set to the destination.

Set $i = 1$

Assume k is the size of candidature set and use $S_1, S_2, \cdots, S_k$ to represent nodes inside the candidature set of node S

while $i < k$ **do**
    $S_{dist}[i] = dist(S_i, D)$
    $i++$
end while

Step 2: Sort $S_{dist}[i]$ in ascending order

Set $u = 0, j = k$

while $u < k$ **do**
    while $j > u$ **do**
        if $S_{dist}[j-1] > S_{dist}[j]$ **then**
            swap($S_{dist}[j-1]$, $S_{dist}[j]$)
        end if
        $j--$
    end while
    $u++$
end while

Step 3: Use S' to represent the selected inter-nodes from the candidature set of node S and use P to represent the estimated delivery percentage value.

node S forwards message to those selected S' with a certain *TTL* and a packet ID (P_{ID}).

Set $r = 0, v = \lceil k * P \rceil + 1$

while $r < v$ **do**
    $Forward(S, Getnode(S_{dist}[r]), TTL, P_{ID})$[1]
    $r++$
end while

Step 4: Once an inter-node receives the forward packet, it will check the packet ID to see whether this packet has been received or not. If the inter-node does not receive this packet and the TTL is not equal to 0, it will repeat step 1, 2, and 3 to multi-cast this packet to the next available inter-nodes until the destination receives this packet.

References

1. Chaintreau, A., Hui, P., Crowcroft, J., Diot, C., Gass, R., Scott, J.: Impact of human mobility on opportunistic forwarding algorithms. IEEE Trans. Mobile Comput. **6**, 606–620 (2007)
2. Gao, L., Li, M., Zhou, W., Shi, W.: Privacy protected data forwarding in human associated delay tolerant networks. In: Proceedings of IEEE TrustCom, Melbourne, July 2013, pp. 586–593
3. Gao, L., Li, M., Bonti, A., Zhou, W., Yu, S.: M-dimension: multi-characteristics based routing protocol in human associated delay-tolerant networks with improved performance over one dimensional classic models. J. Netw. Comput. Appl. **35**(4), 1285–1296 (2012). Intelligent algorithms for data-centric sensor networks

---

[1]Function *Forward*() and *Getnode*() are created to forward messages and find the crossroading node's IDs.

4. McPherson, M., Lovin, L.S., Cook, J.M.: Birds of a feather: homophily in social networks. Annu. Rev. Sociol. **27**(1), 415–444 (2001)
5. Lazarsfeld, P.F., Merton, R.K.: Friendship as a social process: a substantive and methodological analysis. In: Berger, M., Abel, T., Page, C. (eds.) Freedom and Control in Modern Society, pp. 18–66. Van Nostrand, New York (1954)
6. Milgram, S.: The small world problem. Psychol. Today **2**, 60–67 (1967)
7. Scott, J., Gass, R., Crowcroft, J., Hui, P., Diot, C., Chaintreau, A.: Crawdad data set Cambridge/Haggle (v. 2009-05-29). Downloaded from http://crawdad.cs.dartmouth.edu/cambridge/haggle (2009)
8. Chaintreau, A., Hui, P., Scott, J., Gass, R., Crowcroft, J., Diot, C.: Impact of human mobility on opportunistic forwarding algorithms. IEEE Trans. Mobile Comput. **6**, 606–620 (2007)
9. Clauset, A., Shalizi, C.R., Newman, M.E.J.: Power-law distributions in empirical data. SIAM Rev. **51**, 661–703 (2009)
10. Watts, D.J., Dodds, P.S., Newman, M.E.J.: Identity and search in social networks. Science **296**, 1302–1305 (2002)

Chapter 5
Data Dissemination in Delay Tolerant Networks with Geographic Information

"Unstable Network" is one of the most important features of delay tolerant networks (DTNs), where most of services or data dissemination are processed in a mobile environment. Geographic information, therefore, is an important factor needed to be carefully analyzed to overcome this "unstable" environment. In this chapter, how to use geographic information to improve data dissemination in DTNs is presented and two DTN applications based on geographic information are illustrated. The first one is vehicle based DTNs, where the vehicle is a carrier to disseminate data. The second one is human associated DTN, where the human is the carrier.

5.1 Geographic Location Based Data Dissemination in Vehicle Delay Tolerant Networks

With the development of vehicle communications [1] and usage of vehicle as a carrier for data offloading over DTN [2], vehicle based delay tolerant networks (VDTNs) becomes an emerging research topic. In this section, background and two applications in VDTNs are presented.

5.1.1 Background

Vehicle based delay tolerant networks (VDTNs) use vehicles, such as cars, trains, trams and buses, as mobile nodes in delay tolerant networks environment. Compared with the DTNs, VDTNs work in a highly dynamic network topology and short contact [3]. It turns every participating vehicle into a wireless point to provide data dissemination service. In general, each vehicle has capability to connect with other devices in approximately 100–300 m. As vehicle has a large movement range,

© The Author(s) 2015
L. Gao et al., *Delay Tolerant Networks*, SpringerBriefs in Computer Science,
DOI 10.1007/978-3-319-18108-0_5

almost unlimited power supply and capability to carry large and powerful wireless devices, they can be used as data carrier for remote areas, particular for those remote rural communities and regions, and provide offloading services for cellular networks.

Based on vehicles' movement pattern, they are classified into the following three categories:

- Fixed Route: Vehicles move to the predefined destination with pre-configured routes, such as shuttle bus and train.
- Flexible Route with Fixed Destination: Vehicles move to the predefined destination with different routes, such as taxi and GPS enabled car.
- Expected Route: Vehicles move along a certain road, such as the road between two intersections.

Vehicles used in VDTN are mostly equipped with a navigation system, such as GPS, to identify its location and the route to the destination.

5.1.2 Generic DTN Based Data Dissemination in Vehicle Based Delay Tolerant Networks

GeoSpray [4] is a pure store-carry-forward based multi-cast data dissemination protocol in VDTNs, which utilizes mobility of vehicles and the location information provided by GPS or similar devices. Its main idea is to always forward the packet to next hop which has a shorter geographic distance compared with itself.

GeoSpray's data dissemination is summarized into four steps. Assume vehicle A is in the transmission range of vehicle B. In the first step, vehicle A sends a list of its bundles which had been delivered in the VDTN before. Vehicle B could detect the identification, source and destination, and a hash of the content of each known previous delivered bundle. At the meantime, vehicle B also exchange this information with vehicle A. Then vehicle A deletes any bundles, which had been delivered by vehicle B, in its memory. Finally, vehicle A updates its memory to merge both vehicles' information.

In the second step, vehicle A checks its bundles' destination one by one. If its destination is vehicle B, vehicle A forward this bundle to vehicle B and remove this item from its memory. Meanwhile, vehicle A adds this information to its delivered bundle list.

In the third step, vehicle A sends a list of bundles stored in its memory to vehicle B. In addition of the information exchanged during the first step, the bundle size, time to live, number of copies in memory and the minimum estimated time of delivery (METD) are also exchanged in this step.

In the last step, vehicle A analyzes the information exchanged from vehicle B and forwards those bundles to it, if the vehicle B is closer to destination or has a lower METD value. For each bundle in vehicle A's buffer, if this bundle is not in vehicle

B's buffer and the METD value of vehicle A is bigger than vehicle B's value, vehicle A hands over the half of its copies to vehicle B (if its remaining number of bundles is greater than 1). Otherwise, vehicle A just forwards this bundle to vehicle B and delete it from its buffer.

As VDTN is not stable and has limited bandwidth with short contact durations, it may not possible to transfer all bundles at the same time. Therefore, all bundles are sorted first before the transmission. The sorting criteria is that if the bundle has the lowest METD, it will be delivered first. If two bundles have the same METD value, bundles with a longer remaining TTLs are scheduled to be forward first.

In summary, GeoSpray uses DTN technique to delivery bundle in a hybrid model. It starts with the multi-cast mode at the beginning to spread a limited number of copies to exploit alternative paths. Then it switches to the single-cast mode for improving the delivery success ratio and reducing the end-to-end delay.

5.1.3   Hybrid Geographic Information Based Data Dissemination in Vehicle Based Delay Tolerant Networks

Based on previous application, geographic DTN routing is useful and important in vehicle communication. To further improve the data dissemination performance, GeoDTN+Nav [5] deployed a hybrid geographic information based model, which utilizes the greedy, perimeter and DTN modes.

The basic idea of GeoDTN+Nav is to use greedy mode first to forward packets to destination greedily by choosing a neighbor which has a shorter geographic distance towards the destination among all neighbors. When there is no neighbor closer to the destination than itself, e.g. there is an obstacle in front, the perimeter mode is used.

In perimeter mode, three factors are considered to decide whether it can change back to the greedy mode or switch to the DTN mode. These factors are network disconnections, delivery quality of nodes carrying packets and neighbor' moving direction. To detect the network connection, it calculates the number of hop counts the packet has traveled, which assumes that the network is partitioned when hop count in the perimeter mode is increasing. To find delivery quality of nodes carrying packets, neighbors' navigation information is monitored. For example, a good neighbor who can deliver a packet to destination should have a mobility pattern to bring packets close to destination compared with itself. The neighbor's moving direction is to make sure that those nodes with higher delivery quality are willing to deliver packets with right direction. For example, a bus has a fixed route closer to destination, but it could move away from it. In that case, this bus has a good delivery quality, but a wrong moving direction. Based on these three factors, a "switch score" could be calculated to decide when the delivery mode should change to the DTN mode.

If the perimeter mode also fails, the DTN mode is used to forward this packet in this disconnected network environment, which is similar to the previous application.

5.2   Combination of Time, Social and Geographic in Human Associated Delay Tolerant Networks

Routes or pathways of messages in human associated DTNs depend largely on the social connections of individuals, or in other words, on established connections among people [6]. These social connections can be abstracted into different communities based on geographic location and social interests. As the characteristics and behaviors of individuals in a social network change over time, relationships among them will change as well.

In this section, the relationship among social behaviors, time patterns and geographic information is addressed to discuss how to combine these three factors to improve the routing performance in DTNs.

5.2.1   Background

Chapter 4 illustrates the importance and advantage of combining multiple dimensional social attributes in the routing selection process. However, as mobile nodes in human associated DNTs are carried by humans, human movement can also determine the data forwarding performance. As mentioned in [7], social network groups are constrained by geographic information. For example, during an international conference, people in the same session/workshop may have strong research interests in the same field and therefore have a high probability contacting with each other. McPhersion et al. [8] suggest the most basic source of homophily is geography, where people are more likely to have contact with those closer to us in geographic location. Zipf [9] stated it takes more energy to connect to those who further than those who are readily available. On the other hand, recent work [10, 11] shows certain social behaviors are time-dependent, where the time factor playing an important role in the performance of human associated DTNs. For example, the contact frequency among classmates during the day is much higher than at night. This indicates social routing is more accessible during the day rather than at night. Thus, the time factor should also be taken into consideration in the routing decision.

Furthermore, Barabási et al. in [12] proposed all characters of time, social and geography have strong relationships that influence each other. They demonstrated that human contacts have a high degree of time and geographic regularity, where each individual is characterized by a time-dependent characteristic travel distance and a significant probability to return to a few highly frequented locations. In the following sub-section, a case study based on the MIT Reality Dataset [13, 14] is presented to illustrate why it needs to use time, social and geographic information regarding the improvement of data dissemination performance. A time-varying social contact topology model is then proposed to combine these three factors as an overall forwarding decision making strategy.

5.2.2   Case Study Based on MIT Reality Dataset

In this sub-section, a case study based on the MIT Reality Dataset is presented [15]. The Reality Dataset is a real social trace file [13, 14] conducted from 2004 to 2005 in a 9 month period at the MIT Media Laboratory, which is also another benchmark dataset in DTNs followed by Infocom 06 dataset [16], as summarized in Table 5.1.

This project studied 106 subjects by using mobile phones with several pre-installed applications that recorded data about call logs, Bluetooth device IDs, cell tower IDs, application usage and phone status. As delay tolerant networks belong to mobile ad hoc networks, the Bluetooth trace file in the MIT Reality Mining dataset was used as the evaluation data for DTNs and each Bluetooth scan was treated as one contact between two nodes.

To verify and study this dataset's social behaviors, 10 subjects were randomly selected from 106 subjects and their contact frequency between each other in a 2 month period was recorded to form Table 5.2. Furthermore, their social attributes, such as names (Subject ID), affiliation and hashed Bluetooth MAC address were collected, as shown in Table 5.3.

Originally, if there were no social characteristics in this dataset, every node should have an equal chance of contacting other nodes with similar values in terms of contact frequency. However, evaluation results from Table 5.2 demonstrates the direction of network traffic changes when social concepts are introduced into the network. For example, Subject 16 has 149 contacts with Subject 19, but has no contact with Subject 9. This is because, both Subject 16 and Subject 19 are 1st year graduate students, while Subject 9 is a media lab professor, and all these three subjects have different affiliation backgrounds. Therefore, if social affiliation is taken into consideration, most subjects in the same social category will have a

Table 5.1 Summarization of the MIT Reality dataset

Dataset	MIT Reality
Device	Nokia 6600
Duration in days	246
No. of nodes	106
Total edges	54,667
Maximum geodesic distance	4
Average geodesic distance	1.964305
Graph density	0.17829457

Table 5.2 Communication relationship among ten subjects in the MIT Reality Mining dataset

ID	3	9	16	19	36	47	66	83	96	103
3		3	32	48	0	2	0	0	1	12
9	1		2	3	0	0	0	2	4	1
16	43	0		149	1	0	0	0	1	2
19	72	2	131		0	0	0	0	0	3
36	0	1	1	0		0	0	23	0	5
47	0	0	0	0	0		0	0	0	0
66	0	0	0	0	0	0		0	0	0
83	0	1	1	1	7	0	0		1	459
96	0	0	0	1	0	0	0	2		0
103	0	0	2	0	0	0	0	93	0	

Table 5.3 Ten Subjects from the MIT Reality Mining dataset

Subject ID	Affiliation	Hashed MAC
3	1st Year Graduate Student	61961024887
9	Media Lab Professor	61961078506
16	1st Year Graduate Student	61961024963
19	1st Year Graduate Student	61961024937
36	Media Lab Graduate Student	61961025054
47	Null	61964944370
66	Sloan Business School	61964943982
83	Media Lab Graduate Student	61964972168
96	Media Lab Graduate Student	61964979158
103	Media Lab Graduate Student	413791240838

high contact rate. This proves the Bluetooth data in the MIT reality Dataset clearly has social behaviors embedded.

The MIT Reality Dataset records each node's movement but does not release the geographic information of the movement. This drawback has been overcome by Ficek et al. [17], who revealed the relative location information based on analysing node movement information and their associated cell tower IDs. However, they claim only 46.75 % of all unique cell locations from the MIT Reality Dataset can be retrieved with geographical coordinates, and hence the dataset from MIT is not complete. To remedy this deficiency, only the most active nodes (35 nodes) are considered and their location information is visualized from Google Maps, as shown in Fig. 5.1. The longitude and latitude of corners on this map are (42.31, −71.14), (42.38, −71.14), (42.31, −71.04) and (42.38, −71.04) respectively. For calculation and description purposes, this map is divided into a $10 * 10$ grid and the x-y coordinate system is used to measure each grid. To simplify this, the line on (42.31, −71.14) and (42.31, −71.04) is used to represent the x-axis and the line on (42.31, −71.14) and (42.38, −71.14) is used to represent the y-axis.

In this map, most data communication occurred in a small area with blue points (a red circle is used to highlight these blue points). The two biggest circles are in the same range as the cellular towers located on the side of the MIT campus, where

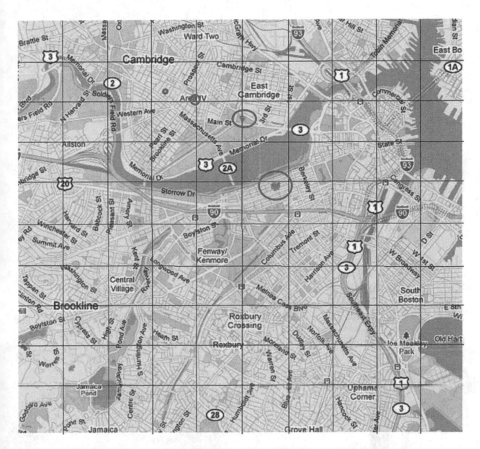

Fig. 5.1 A 2 month contact topology example generated from the MIT Reality dataset

most mobile devices used in the MIT trace file are associated with them. Smaller circles highlight some popular locations, where students and staffs may want go to, such as restaurants and bars. This figure demonstrates that data communication has a strong relationship with its geographic information. Details of the number of contacts in the popular area/grid are illustrated in Fig. 5.2. The area code in this figure is represented by three digital, where the first digital represents the position on the x-axis and the third digital represents the position on the y-axis. This figure also follows the Gaussian distribution, as shown in Fig. 5.3 and its parameters are listed in Table 5.4.

$$f(t) = Ae^{-\frac{(t-u)^2}{2*\sigma^2}} \tag{5.1}$$

Figure 5.4 shows time indexed contact distribution among 106 subjects of the MIT Reality trace file based on 2 months of sample data. This figure illustrates the data forwarding frequency varied differently in time because most contact among

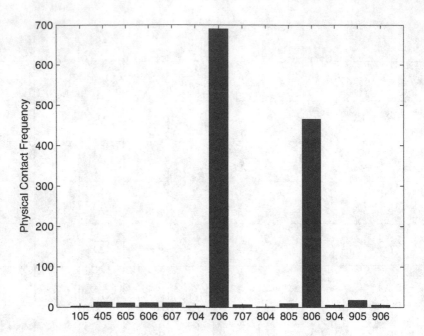

Fig. 5.2 Geographic contact frequency for most popular areas

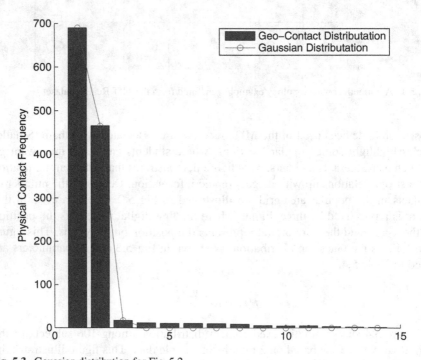

Fig. 5.3 Gaussian distribution for Fig. 5.2

Table 5.4 Gaussian distribution parameters for Fig. 5.3

Parameters	Values
A	837
u	1.364
σ	0.586

Table 5.5 Geographic attribute mapping table for Fig. 5.3

Subject	Geographic area code	Subject	Geographic area code
1	Area 706	8	Area 805
2	Area 806	9	Area 707
3	Area 905	10	Area 904
4	Area 405	11	Area 906
5	Area 606	12	Area 704
6	Area 607	13	Area 105
7	Area 605	14	Area 804

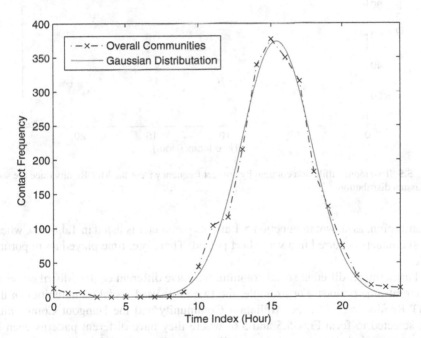

Fig. 5.4 Two month contact frequency sample from the MIT Reality trace file with Gaussian distribution

nodes/people occurred between 14:00 and 18:00. Most participants for the MIT Reality Dataset came from MIT university, where their work from 9 am to 5 pm, Monday to Friday. Therefore, they were supposed to have equal distribution at these time periods. However, the distribution of this figure follows the Gaussian

Parameters	Fig. 5.4	Fig. 5.5	Fig. 5.6
A	373.8	147.4	120.2
u		15.38	15.46	15.45
σ		2.44	2.76	2.42

Table 5.6 Gaussian Distribution Parameters for Figs. 5.4–5.6

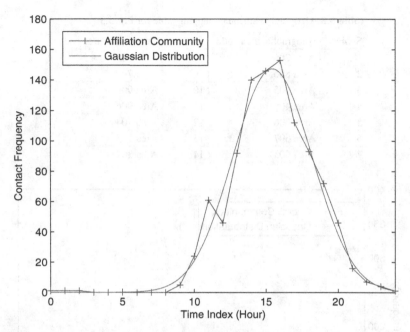

Fig. 5.5 Two month affiliation community contact frequency from the MIT Reality trace file with Gaussian distribution

Distribution, as shown in Function 5.1 and its parameter is listed in Table 5.6, where most contacts occurred in a very short period. Therefore, time played an important role.

Furthermore, different social communities have different contact distributions in different time periods. For example, the two most used social communities in the MIT Reality trace file, the Affiliation Community and the Hangout Community, are selected to form Figs. 5.5 and 5.6, where they have different patterns even in the same period. However, their overall contact distribution follows the Gaussian Distribution and they do have the same trend as Fig. 5.4. Although, during the day, from 9 am to 9 pm, the "Affiliation" community was more active than the "Hangout" community. While, during the night, from 9 pm to 3 am, the "Hangout" community was more active compared to the "Affiliation" community. This is because the Affiliation community is relevant to academic activities that most likely occurred during the day. The Hangout community was more relevant to entertainment activities that most likely occurred during the night. These examples illustrate that

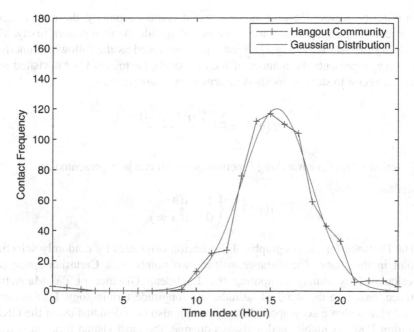

Fig. 5.6 Two month Hangout community contact frequency from the MIT Reality trace file with Gaussian distribution

time pattern plays an important role when discussing contact distribution and each social attribute also has different importance across different time periods.

In summarizing this case study, these three factors, time, social and geography, are important for the routing and data forwarding performance improvement. Hence, to further improve their data forwarding performance of human associated DTNs, these three factors are needed to be combined for an overall evaluation. In the following section, a time-varying social contact topology (TV-SCT) model is proposed to meet the above requirements.

5.2.3   Time-Varying Social Contact Topology Model in Delay Tolerant Networks

As the case study shows above, time, social and geography are very important for human associated DTNs. Therefore, to combine time patterns, social characteristics and geographic information, a time varying social contact topology (TV-SCT) model is presented. This model is used to coordinate these factors into a single system and generate an overall delivery probability to form the critical forward selection.

In order to achieve the time varying social contact topology, the model needs to consider the geographic topology changes alongside the time pattern firstly. The probability of User a visiting Location l is represented as the following function, where $n(a)$ represents the number of locations/cellular towers User a visited and $L_i(a)$ is a vector to store all location information of user a.

$$PL(a,l) = \frac{\sum_{i=1}^{n(a)} \eta(l, L_i(a))}{n(a)}$$

Function $\eta(x,y)$ is a matching function, where it can be represented as:

$$\eta(x,y) = \begin{cases} 1 & \text{if } x = y \\ 0 & \text{if } x \neq y \end{cases}$$

The Distance for the geographical dimension originates by randomly selecting a point in the space. The distance within two points on a Cartesian space can be calculated by simply computing the Euclidean Distance or the Manhattan Distance, based on the standard latitude and longitude terminology. Furthermore, the distance within a geographic dimension can also be calculated using the GRID algorithm [18]. As mobile nodes always change, the most visited location is used to obtain their relevant geographic distance. For User a, its most often visited location is:

$$ML(a) = argmax_{l \in Loc} PL(a,l),$$

where Loc represents all locations User a visited. The relevant geographic distance between two users can be represented as

$$D(a,b) = dist(ML(a), ML(b)),$$

where dist(a, b) is

$$dist(a,b) = \sqrt{(a_x - b_x)^2 + (a_y - b_y)^2}.$$

Therefore, the probability of Users a and b appearing at the same location during the same time period is

$$PT(a,b) = \frac{\sum_{i=1}^{n(a)} \sum_{j=1}^{n(b)} H(\Delta T - T_i(a) - T_j(b)) \eta(L_i(a), L_j(b))}{\sum_{i=1}^{n(a)} \sum_{j=1}^{n(b)} H(\Delta T - T_i(a) - T_j(b))} \tag{5.2}$$

where H(x) is the Heaviside Step Function [19], and ΔT is set to 1 h in default. Function H(x) is defined as

$$H(x) = \begin{cases} 0 & x < 0 \\ \frac{1}{2} & x = 0 \\ 1 & x > 0 \end{cases} \tag{5.3}$$

As mentioned in Chap. 4, users have different social attributes, such as Father, Swimmer and Lecturer. However, it is very hard to detect and record which social attribute appears during a certain temporal, especially if these attributes were recorded in a dataset. To overcome this, an overall social attributes similarity between the source node and target node is calculated. The probability that User a contacts another user with the same social attribute s is:

$$PS(a,s) = \frac{\sum_{i=1}^{n'(a)} \eta(s, S_i(a))}{n'(a)},$$

where $n'(a)$ represents the number of contacts of user a has with another user when both of them share the same attribute and $S_i(a)$ represents a vector storing the social attributes of user a. Therefore, the most likely social attribute for user a is:

$$MS(a) = argmax_{l \in Soc} PS(a,s)$$

To capture the social similarity of user a and user b, the cosine similarity is deployed and the two users' social similarity is represented as

$$Sim(a,b) = \frac{\sum_{s \in Soc} PS(a,s) \times PS(b,s)}{\sqrt{\sum_{s \in Soc} PS(a,s)^2} \times \sqrt{\sum_{s \in Soc} PS(b,s)^2}}$$

With social similarity and time embedded geographic distance of source and destination nodes, the overall time varying social contact topology prediction probability of User a and User b is:

$$Pre(a,b) = PT(a,b) \times Sim(a,b) \tag{5.4}$$

With this prediction probability function, multi-cast data forwarding can be conducted. The source node selects the suitable top N inter-nodes from its neighbor set to forward packets.

The source node uses Function 5.4 to generate all prediction probabilities to the destination node for each of nodes in its neighbor set. An estimate delivery percentage value P is used to select the Top $N = \lceil k * P \rceil$ inter-node(s) to forward to, where there is at least one node selected. With a certain TTL, each packet is issued a unique packet ID by a source node. Once an inter-node receives a forwarded packet, it checks the packet ID to see whether this packet has been received or not. If not received and the *TTL* is not equal to 0, it changes the value of *TTL* to *TTL* − 1 and repeats the above steps to multi-cast this packet to the next available inter-nodes until the destination node receives this packet.

5.3   Summary

In this chapter, a real social trace file, the MIT Reality Mining Dataset, was presented as a case study to study its time, social and geography characteristics. It demonstrated the social behaviour of nodes has a strong time influence, where the importance of social attributes can vary with different time patterns. For example, the Affiliation attribute is important during the day, but the Hangout attribute is more important at the night compared with the Affiliation attribute. Another factor that needs to be considered is geographic information, where nodes in the same location for most of the time may share some similar social interests. In addition, if nodes have similar social interests, they may be physically located in a close area. In summary, these three factors play a vital role in human associated DTNs.

Furthermore, a time varying social contact topology model is proposed to combine social attributes, time patterns and geographic information. This model analyses the social similarity of nodes along with the geographic information bounded with time factors to generate an overall delivery probability for each node to the destination. Based on this delivery probability, a multicast data forwarding algorithm is deployed to efficiently deliver packets.

References

1. Luan, T.H., Cai, L.X., Chen, J., Shen, X., Bai, F.: Engineering a distributed infrastructure for large-scale cost-effective content dissemination over urban vehicular networks. IEEE Trans. Veh. Technol. **63**(3), 1419–1435 (2014)
2. Li, Z., Liu, Y., Zhu, H., Sun, L.: Coff: contact duration aware cellular traffic offloading over DTN. IEEE Trans. Veh. Technol. (2014). 10.1109/TVT.2014.2381220
3. Burgess, J., Gallagher, B., Jensen, D., Levine, B.N.: MaxProp: routing for vehicle-based disruption-tolerant networks. In: Proceedings of IEEE INFOCOM, Barcelona (2006)
4. Soares, V.N., Rodrigues, J.J., Farahmand, F.: Geospray: a geographic routing protocol for vehicular delay-tolerant networks. Inf. Fusion **15**(0), 102–113 (2014). Special issue: resource constrained networks
5. Cheng, P.-C., Lee, K.C., Gerla, M., Härri, J.: GeoDTN+Nav: geographic DTN routing with navigator prediction for urban vehicular environments. Mobile Netw. Appl. **15**(1), 61–82 (2010)
6. Liang, X., Li, X., Luan, T.H., Lu, R., Lin, X., Shen, X.: Morality-driven data forwarding with privacy preservation in mobile social networks. IEEE Trans. Veh. Technol. **61**(7), 3209–3222 (2012)
7. Onnela, J.-P., Arbesman, S., González, M.C., Barabási, A.-L., Christakis, N.A.: Geographic constraints on social network groups. PLoS One 6(4), e16939 (2011). http://journals.plos.org/plosone/article?id=10.1371/journal.pone.0016939

8. McPherson, M., Lovin, L.S., Cook, J.M.: Birds of a feather: homophily in social networks. Annu. Rev. Sociol. **27**(1), 415–444 (2001)
9. Zipf, G.K.: Human Behavior and the Principle of Least Effort. Addison-Wesley, Reading (1949)
10. Gao, W., Cao, G., La Porta, T., Han, J.: On exploiting transient social contact patterns for data forwarding in delay-tolerant networks. IEEE Trans. Mobile Comput. **12**, 151–165 (2013)
11. Huang, M., Chen, S., Zhu, Y., Wang, Y.: Topology control for time-evolving and predictable delay-tolerant networks. IEEE Trans. Comput. (to appear)
12. Gonzalez, M.C., Hidalgo, C.A., Barabasi, A.-L.: Understanding individual human mobility patterns. Nature **453**, 779–782 (2008)
13. Eagle, N., Pentland, A., Lazer, D.: Inferring social network structure using mobile phone data. PNAS **106**, 15274–15278 (2007)
14. Eagle, N., Pentland, A.S., Lazer, D.: Mobile phone data for inferring social network structure. Soc. Comput. Behav. Model. Predict. **36**, 79–88 (2008)
15. Gao, L., Li, M., Bonti, A., Zhou, W., Yu, S.: Multidimensional routing protocol in human-associated delay-tolerant networks. IEEE Trans. Mobile Comput. **12**, 2132–2144 (2013)
16. Scott, J., Gass, R., Crowcroft, J., Hui, P., Diot, C., Chaintreau, A.: Crawdad data set Cambridge/Haggle (v. 2009-05-29). Downloaded from http://crawdad.cs.dartmouth.edu/cambridge/haggle (2009)
17. Ficek, M., Kencl, L.: Spatial extension of the reality mining dataset. In: Proceedings of IEEE MASS, San Francisco (2010)
18. Liao, W., Sheu, J., Tseng, Y.: Grid: a fully location-aware routing protocol for mobile ad hoc networks. Telecommun. Syst. **18**(1), 37–60 (2001)
19. Legua, M.P., Morales, I., Sánchez Ruiz, L.M.: The heaviside step function and MATLAB. In: Proceedings of the International Conference on Computational Science and Its Applications, Part I (ICCSA '08), Singapore, pp. 1212–1221. Springer, Berlin/Heidelberg (2008)

Chapter 6
Privacy Protected Routing in Delay Tolerant Networks

As illustrated in Chap. 3, the most recent research on human associated DTNs uses social information to improve performance, thus demonstrating social characteristics are important factors to improve performance. However, little has been done concerning the privacy issue in DTNs. With increasing use of social information, more and more social characteristics related data is disclosed without permission. Due to the privacy issue, most trace providers do not want to release data with meaningful social information, that may be used to identify the data entered especially for a large dataset. For example, AOL published a query logs but quickly removed it due to the re-identification issues addressed in [1]. Consequently, researchers are trying to find ways to anonymize personal information while maintaining network functioning.

Anonymization [2, 3] is the most useful approach that seeks to anonymize the identity and sensitive data. In datasets, all explicit identifiers such as name and ID number must be removed. But this is not enough. Sweeney [4] discussed a privacy threat to a published health care dataset, from which a former governor of the state of Massachusetts had been re-identified. Furthermore, after linking the medical database with a public voter list, 87 % of individuals could be re-identified with combination of zip code, date of birth, and gender.

Attributes semantics should be reserved in some scenarios, such as data mining or information retrieval. But this is not necessary in DTNs scenario. In DTNs, discovering relationships among the features is more important compared to understanding the meanings of the attributes, and therefore, all attributes can be anonymized. As the main focus of this section is on how to utilize this anonymized data, all attributes are simply anonymized with codes, and values in each attribute are changed to number, which erases the semantics of all attributes in the database. Accordingly, a coding dataset without any meaningful information is generated. An adversary can neither link this dataset with others nor use background information to de-anonymize it [5]. Therefore, social trace providers can safely publish their datasets without violating privacy.

© The Author(s) 2015 69
L. Gao et al., *Delay Tolerant Networks*, SpringerBriefs in Computer Science,
DOI 10.1007/978-3-319-18108-0_6

The key nodes in a network play a critical role in the data forwarding process. Many traditional key node selection algorithms utilize the characters of the physical topology to find key nodes. However, they are hardly successful in mobile social networks due to the mobile nature of humans and the human-to-human effect on the network.

In this chapter, a privacy protected data forwarding model is proposed, using Shannon entropy to quantify the importance level of the anonymous attributes and use the socially aware Kcore decomposition algorithm (S-Kcore) to find the right node to forward to. By using entropy, the most useful social characteristics can be identified and their associated weights can also be assigned. With these weighted anonymized social characteristics, S-kcore is deployed to find the best key node for unicast forwarding.

6.1 Background

6.1.1 Data Privacy

To achieve a better performance in human associated delay tolerant networks, more researchers are using real social trace files to study social behaviors. Some publishers simply replace the original identifiers with some random numbers, however this is not enough, as the published dataset may contain other information that can potentially be linked to individuals. As Samarati [6] illustrated in Table 6.1, the released medical data had suppressed the social security number and name, and therefore the identifier for each individual could not be disclosed. However, the attribute data, such as ZIP, DoB, and Race, may also have appeared in other resources, as shown in the voter list of Table 6.2. As these two tables share some similar attributes, it is possible to combine these to attack the data's privacy. For example, in Table 6.1, there is only one female with the ZIP code 94142 and born on 9/15/61. With this information, this female's name, address and medical record can be re-identified.

The attribute itself does not uniquely identify an individual, but their combination, called quasi-identifier (QID) [7], often falls into a unique or small number of records. Based on the QID concept, Sweeny [8] presented a new privacy model, k-anonymity, which ensures every record has at least k-1 records with the same QID value in database. This protects the privacy of the data by guaranteeing that even if external information is linked to the released database, each released record is directly related to at least other k individuals. The key of k-anonymity is to form equivalent groups through generalization or suppression so every individual is indistinguishable in a group.

Another anonymous data method mentioned by Iyengar [9] is to transform the data using generalizations and suppressions [10]. For example, the DoB attribute is an important factor when analyzing medical records. This attribute can be replaced

Table 6.1 Medical data with anonymous identifiers from [6]

SSN	Name	Race	Date of birth	Sex	ZIP	Marital status	Health problem
		Asian	09/27/64	Female	94139	Divorced	Hypertension
		Asian	09/30/64	Female	94139	Divorced	Obesity
		Asian	04/18/64	Male	94139	Married	Chest pain
		Asian	04/15/64	Male	94139	Married	Obesity
		Black	03/13/63	Male	94138	Married	Hypertension
		Black	03/18/63	Male	94138	Married	Shortness of breath
		Black	09/13/64	Female	94141	Married	Shortness of breath
		Black	09/07/64	Female	94141	Married	Obesity
		White	05/14/61	Male	94138	Single	Chest pain
		White	05/08/61	Male	94138	Single	Obesity
		White	09/15/61	Female	94142	Widow	Shortness of breath

Table 6.2 Re-identifying anonymous data by linking to the public voter list from [6]

Name	Address	City	ZIP	DOB	Sex	Party	...
...	...	...	...	...	...	...	...
Sue J. Carlson	900 Market Street	San Francisco	94142	9/15/61	Female	Democrat	Obesity
...	...	...	...	...	...	...	...

by the year of birth, instead of the day and month. This loss of specificity makes the identification process harder. However, if there is too much generalization, the transformed data may useless.

6.1.2 Key Node Selection

Key nodes are the nodes in the graph network [11] more "central" than others. For example, in Fig. 6.1 the middle node has three advantages to being selected as the key node over other nodes: it has more links, it can reach all other nodes quicker, and it controls the flow between other nodes.

The popular key node selection algorithm is the K-core decomposition algorithm [12]. The idea of the K-core algorithm is to select the key nodes by recursively removing all vertexes with a degree less than k, until all vertexes in the remaining graph have a degree at least k, where k is defined as the number of links. The purpose of the K-core decomposition algorithm is to identify particular subsets of the networks.

To help understand the K-core in details, the definition of K-core from Batagelj [13] is used. Assume, there is a graph $A = (V, L)$ of $|V| = n$ vertices and $|L| = l$ edges. A sub graph $H = (C, L|C)$ induced by the set $C \subseteq V$ is a K-core or a core of order k if $v \in C$ and $degree_H(v) \geq k$, then H is the maximum sub-graph

Fig. 6.1 The middle node is
the key node

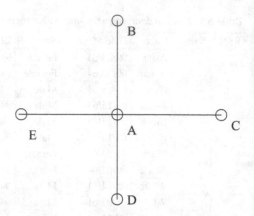

with this property. The K-core of a graph can be obtained by recursively removing all nodes with a degree less than k until all nodes in the remaining graph have a degree of at least k. A vertex i has coreness k if it belongs to the K-core but not belongs to the $(k+1)$-core. Every node in a connected graph belongs to the 1-core, so its coreness is 1, which is also the minimum coreness of a vertex. All vertexes of degree $d \leq 2$ are recursively cut out of 1-core to 2-core, and the maximum vertex coreness is such that the K_{max}-core is not empty, but the $(K_{max}+1)$-core is, and then named the graph coreness. The vertex coreness denotes how "deep in the core" the vertex is, as mentioned by Myers [14].

6.2 Entropy Based Anonymous Data Forwarding in Delay Tolerant Networks

In this section, an entropy based anonymous data forwarding model in human associated delay tolerant networks [15] is illustrated. This model has three main parts: an entropy based anonymous attributes detection algorithm, a social aware K-core decomposition algorithm, and an anonymous data forwarding model.

6.2.1 Entropy Based Anonymous Attribute Detection

In a human associated DTNs scenario, attribute semantics could be anonymized as they cannot manually decide which attribute is more important based on the human senses. Therefore, all attributes are anonymized as codes, and values are anonymized as numbers. As the example shown in Chap. 4 demonstrates, in the university school structure there are five variable values: Head of school, Student, Unit chair, Lecture, and Tutor. In typical anonymous methods, these values might be generalized as teacher and student. In our anonymous method, however, occupation is coded as letter A, and five values are replaced by 1, 2, 3, 4, and 5. By using this method, the true meaning of the code cannot be retrieved and the privacy of individuals can be highly protected.

When an anonymity trace file is provided, the first thing is to find relationships among each feature where it needs to differentiate useful social characteristics from this unpredictable dataset. For example, assume there are ten researchers coming from eight different universities and they are either lecturers or professors. In this example, there are two dimensions: university location and affiliation. Based on our normal understanding, university location is more important than the affiliation. This is because there is a higher opportunity for two people to belong to the same affiliation while they may have lower chance of knowing each other. However, if two people belong to the same university, they most likely know each other.

In order to quantify these relationships, entropy is used to measure their uncertainties. This means if a metric has a higher uncertainty, it is better than a lower uncertainty in allocating weight. Assume there are m anonymous dimensions for node S and for each dimension it has n possible values, $x_1, x_2, \cdots, x_n$.

For the MIT Reality and Infocom 06 datasets, entropy values for each of the anonymized attributes are listed in Tables 6.3 and 6.4. Attributes for each anonymous code are listed in Table 6.5.

Table 6.3 Entropy of the MIT Reality Mining dataset

A	2.4289598
B	4.691676
C	3.4489806
D	1.206241
E	1.4099114
F	6.022368
G	0.84963125

Table 6.4 Entropy of the Infocom 06 dataset

A	4.160289
B	4.9075212
C	3.38649
D	4.5660243
E	1.3562925
F	4.467997
G	3.550113

Table 6.5 Mapping between anonymity symbols and their corresponding characters in associated datasets

Anonymity symbol	Symbol in MIT dataset	Symbol in Infocom dataset
A	Affiliation	Nationality
B	Group	Studies
C	Office	Language
D	Provider	Affiliation
E	Community	Position
F	Mac Address	City
G	Phone Plan	Country

With the Shannon Entropy [16], the uncertainty value of a certain dimension D_m^s for node S can be generated as:

$$H(D_m^s) = -\sum_{i=1}^{n} p(x_i) \log_2 p(x_i)$$

The $p(x_i)$ denotes the probability mass function of the outcome of x_i, which can be simplified as:

$$p(x_i) = \frac{x_i}{\sum_{j=1}^{n} x_j}$$

Based on these generated uncertainty values, weight factors can be assigned to each dimension automatically. For Node j, its weight factor is:

$$w_j = \frac{D(F_j)}{\sum_{i=1}^{m} D(F_i)}$$

At this stage, the useful social attributes are differentiated from the anonymized trace file, and based on the different attribute importance levels, the S-Kcore algorithm is applied to find the key nodes to forward to.

6.2.2 Social Aware Kcore Decomposition Algorithm

As unicast forwarding is used in this model, how to select the best inter-node to make sure the forwarding node is alive becomes a critical issue. The number of powerful nodes in a network are few, but they do play an important role. For example in Fig. 6.2, it shows the number of contacts among popular nodes for a 1 month data sample collected from the MIT Reality Minning dataset. This example demonstrates two nodes, Node 4 and 60, take around 45 % of the total message forwarding.

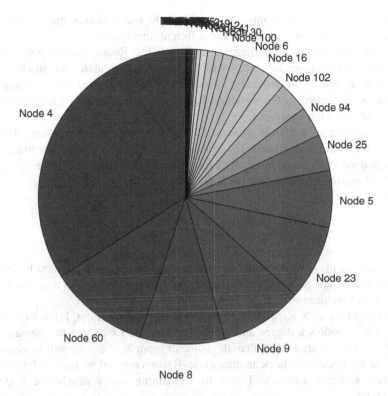

Fig. 6.2 Example of popular nodes in MIT Reality Mining dataset

Therefore, in this case, the most efficient way of data forwarding is to send messages to those "Key" nodes.

Traditional key node selection algorithms [12], however, only focus on physical connections and do not consider the social status of mobile nodes. This consideration results in poor performance in human associated DTNs for the following two reasons. First, the physical connections in human associated DTNs are not reliable due to the mobile nature of the network. Thus, the node degree k related to the physical connection is less trusted compared to a static network, such as a wired network. Second, some nodes in the network have fewer links but are still important because they are carried by important people.

To address these limitations, a socially aware Kcore key nodes selection algorithm, S-Kcore, is proposed in this section, which modifies the well-known Kcore algorithm [12]. The S-Kcore algorithm is designed to exploit the nodes' social features to improve performance in data communication.

The Social Aware Kcore Decomposition Algorithm (S-Kcore) is designed to find key nodes in a social network environment with weighted social attributes. In S-Kcore, the socializing network structure is built on top of the geographic topology. Nodes can also be grouped into different clusters according to their social characters, such as occupation, kinship, interests etc. Thus, the foundation of the S-Kcore idea is a network can be modeled from these different dimensions.

The idea of S-Kcore algorithm is based on the Kcore decomposition algorithm [12] that removes all nodes with a degree less than K. We modified the Kcore decomposition algorithm by introducing a new k-degree computation. The k-degree is re-defined as the weighted function of the number of social contacts using anonymous attributes.

The K-degree of a mobile node in DTNs is defined as the number of social contacts with their neighbors. To represent the affect of social attributes, the importance of each anonymous attribute should be combined, as generated by Eq. 6.1. Therefore, the K-degree for Node n is:

$$K_n = \left\lceil \sum_{i \in A} w_i \times C_i^n \right\rceil, \tag{6.1}$$

where A represents all attributes Node n has and C_i represents all contacts for Node n in its attribute i. The above equation defines the K-degree value as a rounded integer of the overall contacts with different weighted social attributes.

The main loop of S-Kcore is composed by Algorithm 2. First, Eq. 6.1 is applied to calculate a node's k-degree and using this generated k-degree to compare with input K. If the k-degree is less than the value of input K, the node will be removed, otherwise the node will check another node. Recursively, all nodes will be checked with their k-degree values and only the remaining nodes satisfy the K-degree requirement.

Algorithm 2 S-Kcore decomposition algorithm

---
input: K value
$N(s)$ is empty with marking k-degree
while $\forall i \in N(s)$ and it is not marked by k-degree **do**
    select node n_i in $N(s)$
    calculate k-degree using Eq. 6.1
    mark the node n_i with the k-degree
    if k-degree $< K$ **then**
        remove the node n_i
    else
        select another node
    end if
end while

6.2.3 Anonymous Data Forwarding in Delay Tolerant Networks

Data forwarding is the key issue in delay tolerant networks and in order to improve the delivery success ratio, many routing protocols use the multi-cast technique, such as the one used in Chaps. 4 and 5. However, using multi-cast technique increases the rate of redundant messages which have been copied multiple times. This is not good for DTNs as most are energy and bandwidth constrained. In this section, a unicast data forwarding model is proposed.

In a unicast routing environment, there is only one copy of a message sent from the source and this becomes the only copy in the network. Therefore, it is important to select the next inter-node appropriately. Otherwise, if the selected node cannot forward the message to the destination or the next available inter-node, the destination node is no longer able to receive this message.

The unicast model in this section takes both the current network situation and history contact information into consideration. For example, a source node (Node S) wants to send a message (M) to a destination node (Node D). Node S first checks whether Node D is in its Current-Neighbor set or not. If it does have Node D in its neighbor set, Node S directly forwards message M to Node D, and the forwarding process is finished. If Node D does not belong to the Neighbor set of Node S, Node S checks its Previous-Neighbor set to decide the waiting period ΔT and then applies the S-Kclose technique to find the most popular node as the next forwarding target. Details of this forwarding algorithm can be found in Algorithm 3.

Algorithm 3 Unicast Anonymous data forwarding

Step 1: Node S checks its Current-Neighbor set (CN_S) to see whether Node D is directly connected to Node S or not.

if Node $D \in CN_S$ **then**

    Node S directly forward message M to Node D and the forwarding process is finished.

else

    Move to **Step 2**.

end if

Step 2: Node S checks its Previous-Neighbor set (PN_S) to see whether Node D had contacted itself before or not.

if Node $D \in PN_S$ **then**

    Node S checke how many times Node D contacts with itself, the number of contacts with all nodes (including Node D), and the overall time to contacts these nodes.

$$\Delta T = \frac{No.ofcontactwithNodeD}{OverallContactFrequency} \times OverallTime \tag{6.2}$$

else

    The waiting period ΔT is set up as 0 and move to **Step 3**

end if

Step 3: Node S waits a time period ΔT and uses S-Kcore function to figure out the most popular node (Node P) in its Current-Neighbor set and forward to.

Step 4: Node P repeats these steps until the message M is sent to Node D

6.3   Summary

In this chapter, an anonymous data forwarding model was proposed. This model does not require the use of meaningful data to analyze the social behaviors or relationships of nodes. Instead, an Entropy based anonymous data detection model was proposed to measure how important each attribute and their overall relationships were. Based on these relationships and their importance, each attribute was assigned a weight factor to identify its position in this anonymous dataset.

Furthermore, a social aware K-core (S-kcore) algorithm was designed to combine these generated weight factors to detect the key node. Although this key node may not be physically located in the neighbor area of the source node, the S-kcore algorithm was based on social attributes and their associated weight factor where it is not be affected by the physical movement of nodes.

Another advantage of the proposed model is the unicast forwarding model. As the network topology of human associated delay tolerant networks is unpredictable, the concurrent most popular node is selected by using the S-kcore algorithm to conduct unicast forwarding to reduce those duplicated copies.

References

1. Barbaro, M., Zeller, T.: A face is exposed for AOL searcher no. 4417749. New York Times **9**(2008) (2006)
2. Bayardo, R., Agrawal, R.: Data privacy through optimal k-anonymization. In: Proceedings of IEEE ICDE, Tokyo, pp. 217–228. IEEE (2005)
3. Lu, R., Li, X., Luan, T.H., Liang, X., Shen, X.: Pseudonym changing at social spots: an effective strategy for location privacy in VANETs. IEEE Trans. Veh. Technol. **61**(1), 86–96 (2012)
4. Sweeney, L.: k-anonymity: a model for protecting privacy. Int. J. Uncertain. Fuzziness Knowledgebased Syst. **10**(5), 557–570 (2002)
5. Liang, X., Zhang, K., Shen, X., Lin, X.: Security and privacy in mobile social networks: challenges and solutions. IEEE Wirel. Commun. **21**(1), 33–41 (2014)
6. Samarati, P.: Protecting respondents identities in microdata release. IEEE Trans. Knowledge Data Eng. **13**, 1010–1027 (2001)
7. Wang, K., Xu, Y., Fu, A.-C., Wong, R.-W.: FF-anonymity: when quasi-identifiers are missing. In: Proceedings of IEEE ICDE, Shanghai, pp. 1136–1139 (2009)
8. Sweney, L.: Achieving k-anonymity privacy protection using generalization and suppression. Int. J. Uncertain. Fuzziness Knowledge-Based Syst. **10**(5), 571–588 (2002)
9. Iyengar, V.S.: Transforming data to satisfy privacy constraints. In: Proceedings of ACM SIGKDD, Edmonton, pp. 279–288, ACM, New York (2002)
10. Sweeney, L.: Datafly: a system for providing anonymity in medical data. In: Proceedings of IFIP TC11 WG11.3 Eleventh International Conference on Database Security XI: Status and Prospects, pp. 356–381. Chapman & Hall, London, Lake Tahoe, California (1998)
11. Harary, F.: Graph Theory. Addison Wesley, Reading (1969)
12. Alvarez-Hamelin, J.I., Dall'Asta, L., Barrat, A., Vespignani, A.: k-core decomposition: a tool for the visualization of large scale networks. CoRR, abs/cs/0504107 (2005)
13. Batagelj, V., Zaversnik, M.: Generalized cores. CoRR, cs.DS/0202039 (2002)

14. Myers, C.R.: Software systems as complex networks: structure, function, and evolvability of software collaboration graphs. Phys. Rev. E **68**, 046116 (2003)
15. Gao, L., Li, M., Zhu, T., Bonti, A., Zhou, W., Yu, S.: AMDD: exploring entropy based anonymous multi-dimensional data detection for network optimization in human associated DTNs. In: Proceedings of IEEE TrustCom, Liverpool, pp. 1245–1250 (2012)
16. Shannon, C.E.: A mathematical theory of communication. Bell Syst. Tech. J. **27**(3), 379–423 (1948)

Chapter 7
Conclusions and Future Work

This chapter reviews the whole monograph and summarizes its main contributions. Furthermore, potential research topics and their implementation are discussed for future work.

7.1 Monograph Summary

Mobility and long latency are two main characteristics of delay tolerant networks (DTNs) that make traditional routing protocols for mobile ad hoc networks are unsuitable. Most routing protocols for mobile ad hoc networks need to have at least one available end-to-end path between the source and the destination before transmission is conducted. The store-carry-forward technique is used for data forwarding in DTNs. If there was no connection available at a particular time, the source node needs to store and carry this message until another node is encountered and makes a decision whether to forward the message to this node.

When mobile nodes are associated with humans to form a human associated DTNs, they inevitably demonstrate certain social aspects associated with human-to-human communication. More specifically, by collecting and analyzing human social behaviors, the data forwarding performance in DTNs can be improved. Traditional routing protocols in DTNs mainly focus on analyzing the physical topology movements, such as measuring the distance between two nodes in their geographic dimension. However, the geographic topology in the DTNs can be unstable and unpredictable that the social characters of mobile nodes were relatively stable for a fixed period of time. Analyzing the social relationships among nodes becomes a valuable way to improve routing performance.

The data forwarding issue is a fundamental problem for delay tolerant networks. To deliver messages in these networks, existing algorithms use flooding or partial flooding with a probability formulation. There is high probability that any flooding

© The Author(s) 2015

L. Gao et al., *Delay Tolerant Networks*, SpringerBriefs in Computer Science,
DOI 10.1007/978-3-319-18108-0_7

may cause network congestion or high interference, and consume many more resources for processing and switching operations. Context aware routing protocols utilize context information such as history, battery status and changes to the rate of connectivity to compute delivery probabilities. One reason of the above routing mechanisms use flooding or partial flooding is due to a lack of routing information. The geographic information has been proved as an important factor in DTNs, particular in vehicle based DTNs.

Routes or pathways for messages in human associated DTNs depend largely on the social connections of individuals, or in other words, on the established connections among people. These social connections can be abstracted into different communities based on geographic location and social interests. As the characteristics and behaviors of individuals in a social network change over time, relationships among them will change as well. Therefore, these social activities in human associated DTNs are bounded with a time pattern. To obtain a better view of the characteristics of human associated DTNs, social attributes, geographic information and a time pattern are need to be combined as an overall forwarding selection criteria.

Privacy protection is another important factor to be addressed. As the data forwarding issue becomes more and more important in human associated delay tolerant networks, many researchers using real social trace files to study the characteristics of mobile nodes. However, these datasets are sensitive, with potential attackers use the provided social attributes to retrieve the identity of a subject. Therefore, data holders may not want to release their datasets.

Attribute semantics should be reserved in certain scenarios, such as data mining or information retrieval, however it is not necessary in a human associated DTNs scenario. As the information needed for the routing decision making process is only based on relationships among features, all attributes can be anonymized. If both the identification of nodes and their associated attributes are anonymous, the privacy disclosure issue can be prevented and the social trace files can also be published.

7.2   Monograph Contribution

There are five major contributions of this monograph. These are the development of applications of DTNs, routing protocols for DTNs, multiple dimensional social attributes based data forwarding models in DTNs, the combination of geographic information, social attributes and time patterns with weight factors to further improve the forwarding performance in DTNs, and the privacy protected anonymous data detection and forwarding model in DTNs.

Chapter 2 reviewed applications of delay tolerant networks. Applications based on traditional networks need to set up a complete path between source and destination, while they cannot fit into those challenge environments, such as digital communication for rural areas, battlefield communications and disaster rescues. Applications of DTNs overcome these drawbacks and can be widely applied to different carriers, such as human body, vehicle, plane, animal and sensor monitor.

Chapter 3 studied routing protocols based on delay tolerant networks. It first reviewed those epidemic based, which are also fundamental, routing protocols with their advantages and disadvantages. Then the improved routing protocols for delay tolerant networks were addressed. They are probability based routing protocols, geographic based routing protocols, social concept based routing protocols and time related routing protocols.

Chapter 4 presented two multiple dimensional data forwarding models in human associated delay-tolerant networks. These were the Triangle Coordinate model and the M-Dimension model. These models take different social characteristics into account from multiple dimensions and demonstrate the importance of each dimension is. The triangle coordinate model combines the three most useful social attributes into a triangle system, where each angle between two axis represents the social relationship between two social attributes. The smaller an angle, the closer their social relationship. For the M-Dimension model, the weighting process and the distance function were deployed to calculate the distances between two nodes in each dimension by combining the nodes with associated weight factors. Based on these social distances, a multi-cast routing selection algorithm was designed, noting the number of available candidatures and the network change speed. Furthermore, a benchmark dataset in this research field, Infocom 06 dataset, was introduced.

Chapter 5 studied the importance of the geographic factor in vehicle based DTNs and their applications. MIT Reality Mining dataset, which is another benchmark dataset in this research field, was illustrated and its results demonstrated the social behaviors of nodes have a strong time influence, and the importance of social attributes can vary with different time patterns. Furthermore, the MIT Reality Mining dataset demonstrated that the social relationships and geographical locations have a strong correlation, where the majority of connections in the same area tend to have similar social backgrounds. A time varying social contact topology model was proposed to combine social attributes, time patterns and geographic information. This model analyzed the social similarity of nodes and the geographic information bounded time factors to generate an overall delivery probability for each node to the destination. Based on this delivery probability, a multicast data forwarding algorithm was proposed to efficiently deliver packets.

Chapter 6 analyzed the requirement of privacy protection in human associated DTNs and proposed an anonymous data forwarding model. This model did not require the use of meaningful data to analyze nodes' social behaviors and relationships. Rather an Entropy based anonymous data detection method was used to automatically detect the most important attribute as the relationships among features were more important compared with understanding meanings of attributes, and therefore all attributes can be anonymous. Consequently, track provider can safely publish their dataset without violating privacy. In this case, if the Entropy value was bigger, the certainty would be lower, thus indicating the greater importance of the attribute. Based on these relationships and importance, each attribute can be assigned a weight factor to identify its position in the anonymous dataset. A social aware K-core algorithm (S-kcore) was also introduced in this chapter to find the key node based on social information, which is a modification of the K-core algorithm. By using S-kcore and weighting factors, key nodes in the network can be selected based on their social relationships instead of geographic information, as location information was restricted because of privacy issues. Furthermore, a unicast forwarding model is proposed to forward the concurrent most popular node, generated by the S-kore algorithm.

7.3 Future Work

In this section, potential future research topics and application based on delay tolerant networks are presented and discussed.

Energy Efficiency: mobile devices are key components in delay tolerant networks. In most of cases, mobile devices have limited battery and they have to always standby to prepare store-carry-forward services. How to improve the energy efficiency of these opportunistic communications given the tradeoff at the moment between energy consummation and data forwarding performance becomes an important research topic. A potential solution is to design and implement a wake up mechanism, where mobile devices only work when they are required to work. This solution would increase the energy efficiency, while maintaining a high level of data forwarding performance [1–3].

Data Dissemination in Fog Computing: Fog computing [4] is a paradigm to extend the cloud service to the "ground", where applications or contents located on far away worldwide cloud servers are now distributed to hundreds of thousands of Fog servers located just in local business premises, on-board or even the outdoor. These Fog servers have process, storage and network transmission capabilities. Mobile users access those Fog servers with just one-hop wireless connection and therefore streaming applications for majority mobile users becomes available even the area they located has a poor Internet coverage. However, the data dissemination from Cloud to every Fog servers are expensive, particularly for those areas without a fast and reliable Internet access. As the majority of those streaming application are video based, such as movies, teleplay and product advertisement, they are not always need to up-to-date and one or two day latency is affordable. Consequently, delay tolerant network technique becomes a valuable solution for data dissemination of those high volume and non-urgent data among Fog servers.

Natural Disaster Rescue: A potential application based on human associated delay tolerant networks is natural disaster rescues, such as rescues after an earthquake. When an earthquake occurs, mobile phones are the most important and probably the only device humans take along and can be used to make contact. However, during a natural disaster most base stations are damaged and cellular networks may no longer work. As most mobile devices are Bluetooth/Wifi enabled, they automatically form a DTN in a small area if a special app is pre-installed. Based on this network, survivors can be detected and located in a small area. Furthermore, with the storing-and-forwarding technique, a potential communication system can be established in these affected areas. How to establish a reliable and quality-of-service guaranteed communication protocol which is resilient to the dynamic rescue environment [5, 6] therefore deserves further study.

References

1. Ismail, M., Zhuang, W., Serpedin, E., Qaraqe, K.: A survey on green mobile networking: From the perspectives of network operators and mobile users. IEEE Commun. Surv. Tutor. (to appear)
2. He, S., Li, X., Chen, J., Cheng, P., Sun, Y., Simplot-Ryl, D.: EMD: Energy-efficient P2P message dissemination in delay-tolerant wireless sensor and actor networks. IEEE J. Sel. Areas Commun. 31(9), 75–84 (2013)
3. Choi, B., Shen, X.: Adaptive asynchronous sleep scheduling protocols for delay tolerant networks. IEEE Trans. Mobile Comput. 10(9), 1283–1296 (2011)
4. Bonomi, F., Milito, R., Zhu, J., Addepalli, S., Fog computing and its role in the internet of things. In: Proceedings of the First Edition of the MCC Workshop on Mobile Cloud Computing (MCC '12), pp. 13–16, ACM, New York, Helsinki, Finland (2012)
5. Abdrabou, A., Zhuang, W.: Statistical QoS routing for IEEE 802.11 multihop ad hoc networks. IEEE Trans. Wirel. Commun. 8(3), 1542–1552 (2009)
6. Fan, J., Chen, J., Du, Y., Wang, P., Sun, Y.: Delque: A socially aware delegation query scheme in delay-tolerant networks. IEEE Trans. Veh. Technol. 60(5), 2181–2193 (2011)

Printed in the United States
By Bookmasters